The Automation Revolution: Building a Safer, More Fulfilling Future with Robots

Sanobar

Contents

Chapter 1

Introduction

In the last decades, robots have taken over several repetitive and dangerous activities from humans. The automation process has been particularly successful in industry, where the controlled environments and the predictable nature of the conditions simplify the operation of robots. Examples of successful industrial automation can be found in factories along the production lines, where robots perform tasks such as assembling, welding, handling and cleaning materials, and in large warehouses, where robots autonomously organize, sort, and stock items.

More recently, the adoption of robots in other areas of activity has been accelerated by important advancements in some core robotic technologies such as sensors and actuators, manipulators, control systems, materials, batteries, and artificial intelligence. Today, robots work alongside human operators in smart factories to speed up processes, help sorting packages and route them towards their shipment destination, and oftentimes even deliver the packages themselves, hold conversations with humans and answer their questions, assist surgeons during delicate operations, autonomously drive vehicles on crowded streets.

Automation is expected to pervade our society even further in the next decades. The ultimate goal of the automation process is to create a safer and more fulfilling society, in which every person can express themselves fully, without the burden of having to perform dangerous, repetitive and alienating tasks. To achieve this ambitious goal, several technological steps are still required. Some tasks are by nature very challenging to automate, as they have to be performed in unstructured/unknown/hazardous environments, where the working conditions are particularly difficult to predict or can change quickly during the operations. These conditions often penalize rigid systems that require centralized infrastructures to work. On the other hand, they promote systems that are flexible, fault tolerant, and able to quickly adapt their behavior to the contingencies they encounter. Swarm robotics (Beni, 2005; Şahin, 2005; Dorigo et al.,

2014; Yang et al., 2018) is an approach to robotics whose goal is to create systems deemed to perform in such conditions. Swarm robotics takes inspiration from collective behaviors of social animals to develop multi-robot systems that, as their natural counterparts, are flexible, robust, and autonomous (Camazine et al., 2001).

In swarm robotics, a mission is entrusted to a large group of robots, the *robot swarm*. A robot swarm is a highly redundant system that operates in an autonomous and self-organized way, without the need of centralized coordination or external infrastructures. A robot swarm comprises a large number of robots with limited capabilities. The interaction of the robots with each other and with the environment engenders emergent properties: collectively, the swarm displays abilities that a single robot does not possess.

Despite their promising potential, designing robot swarms to perform desired missions is particularly challenging because of their distributed and self-organized nature: While the goal mission is set for the robot swarm as a whole, the designer usually works at a lower level by implementing the behavior of the individual robots that compose the swarm. The designer thus implements the individual behavior so that the interactions between all the individuals and their environment will engender a collective behavior capable of performing the desired mission. For the time being, no general methodology exists for the design of robot swarms. Currently, the design is mostly carried out by hand using a trial-and-error process. In recent years, some principled design methods and a few tools have been proposed to support the design of robot swarms for specific classes of missions under specific assumptions (Hamann and Wörn, 2008; Kazadi, 2009; Berman et al., 2011; Brambilla et al., 2014; Reina et al., 2015b; Lopes et al., 2016). A few automatic (and semi-automatic) design methods have been proposed, but they also operate under various assumptions (see Francesca and Birattari (2016) and Bredeche et al. (2018) for in-depth discussions). Both principled and automatic design are not a reality yet, and more research is needed to make them reliable and generally applicable.

Because of the lack of a general design methodology and of the challenges posed by the design of complex collective behaviors, research has focused mainly on simple classes of missions. In particular, the focus has been mostly on the emergence of geometrical/spatial properties and mechanical abilities: for example, aggregating (Gauci et al., 2014b; Silva et al., 2017), covering space (Schwager et al., 2006; Duarte et al., 2016), forming shapes (Rubenstein et al., 2014b; Mathews et al., 2017), moving coordinately (Virágh et al., 2014), overcoming obstacles (O'Grady et al., 2010), transporting objects (Rubenstein et al., 2013), clustering objects (Gauci et al., 2014a), or assembling structures (Werfel et al., 2014). Less work has been devoted to the emergence of simple cognitive abilities: for example, selecting an aggregation area (Halloy et al., 2007; Garnier et al., 2009; Ozdemir et al., 2018), a behavior (Pini et al., 2011; Castello et al., 2016), a foraging source (Gutiérrez et al., 2010; Valentini et al., 2016b), or a path (Schmickl and Crailsheim, 2008; Montes de Oca et al., 2011; Reina et al., 2015b; Scheidler et al., 2016) between (typically two) alternatives.

Some studies have been already devoted to designing swarms that, inspired by mechanisms of division of labor observed in insect societies (Wilson, 1980; Seeley, 1996; Bonabeau et al., 1998), perform multiple tasks transitioning from one to another (Krieger et al., 2000; Nouyan et al., 2009; Schmickl et al., 2011; Duarte et al., 2016). Nonetheless, one assumption that all previous works share is that the tasks that an individual robot must perform for the desired collective ability to emerge was known at design time. Often in these works, the conditions for transitioning from task to task were also known at design time. The designers of these robot swarms could thus devise and hard-code in the robots' behavior the rules that trigger the transition from task to task. If on the one hand these assumptions simplify the design process, on the other hand they limit the autonomy and the flexibility of the resulting robot swarms: Unfortunately, in most real-world missions, the order of task execution is unknown at design time and it is instead necessary to figure it out at operation time.

In this book, we present *task sequencing swarm* (TS-Swarm), a robot swarm that sequences tasks autonomously at operation time. The task-sequencing ability of TS-Swarm is a cognitive ability that allows the robot swarm to operate even if the correct order of execution of the tasks is unknown at design time. TS-Swarm addresses the case in which m tasks must be performed in a specific order (without repetitions) by an individual robot of the swarm. Each task must be performed in a certain area, and the correct order is unknown at design time. The sequence of tasks must be repeated multiple times by the same or by other robots. Consider, as an example, a swarm of fruit-picking robots (Fig. 1.1). Three tasks must be performed in a specific order by an individual robot: Get a crate at the shed, fill the crate with fruit at the orchard, and finally load the crate onto the truck at the yard. The correct sequence must be repeated multiple times to fully load the truck. The robots initially ignore the correct order of execution. They learn collectively from successes and failures; for example, a robot faces a failure if it reaches the orchard without a crate to fill or the truck with an empty one. Each task must be performed in a specific area: shed, orchard, or yard. Robots travel between these areas to perform the tasks. Multiple robots can operate in parallel: at a given moment in time, some get a crate, some fill their crate, some load the truck, and some other travel from area to area.

The characterizing feature of TS-Swarm is that some of the robots position themselves to form a chain that fulfills two functions: 1. to assist the navigation between the relevant areas where tasks are to be performed and 2. to identify/encode the order in

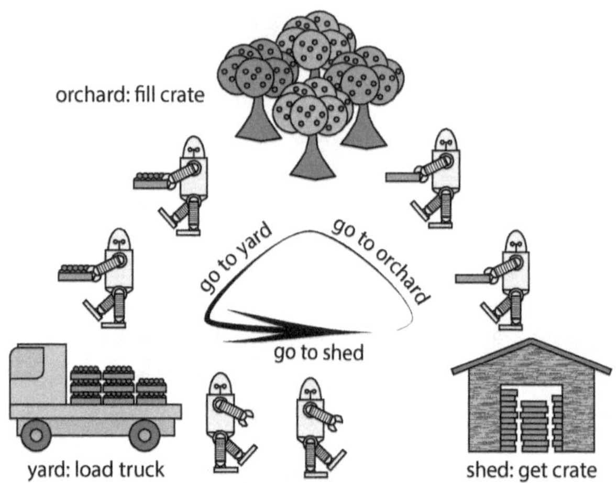

Figure 1.1: **Example of mission that requires sequencing tasks.**

which tasks must be performed. The chain enables robots with limited capabilities to accomplish a complex mission. Individually, the robots of TS-Swarm would be unable to navigate reliably from area to area or to perform the tasks in the correct order. They have a limited range of perception, are unaware of the position of the areas, and are unable to localize themselves in the environment. Moreover, the robots are not programmed to individually sequence tasks by reasoning symbolically on their order of execution.

Chaining has been previously explored in swarm robotics as a means to search the environment and assist navigation (Goss and Deneubourg, 1992; Drogoul and Ferber, 1992; Werger and Matarić, 1996; Nouyan and Dorigo, 2006; Nouyan et al., 2008, 2009; Sperati et al., 2011; Ducatelle et al., 2011a; Dorigo et al., 2013). Nonetheless, to the best of our knowledge, the concept of robot chain has never been associated with path planning nor with planning in general. In this **book**, we acknowledge chaining as a path planning method and we generalize it to planning task sequences. Generally speaking, in the context of swarm robotics, a chain is a group of robots that align in space thus creating a precedence relation: one robot is positioned after the other. So far, in the swarm robotics literature, chains of robots have been conceived as sequences of robots that landmark the physical space and act as waypoints for other robots that need to navigate from one end of the chain to the other. In TS-Swarm, we generalize this picture to include robots that, in a sense, "align" in the abstract space of the tasks, one after the other, creating a precedence relation between the tasks themselves: one task must be performed after the other. These robots, in a sense, "landmark" the abstract space of the tasks and act as logical waypoints for other robots that need to perform the tasks in the order encoded by the chain. The chain that the robots create in TS-Swarm accomplishes the double role of guiding robots in the physical space and in the abstract space of the tasks. It could be considered as a chain that develops in the physical space augmented with the abstract space of the tasks. In the physical subspace, the chain encodes the information needed for navigating from area to area; in the abstract space of the tasks, it encodes the order in which tasks themselves must be performed. By following the information provided by the chain, other robots navigate between the areas where the tasks are to be performed and perform those tasks in the order encoded.

In this **book**, we present two versions of TS-Swarm: Mark I_m and Mark II_m, where m is the number of tasks to be sequenced. Mark I_m assumes that a robot receives negative feedback as soon as it performs a task in an incorrect order and positive feedback otherwise. A robot receives feedback in the sense that, after performing a task, it becomes immediately aware of whether the task was performed in the correct order or not. After demonstrating the task-sequencing ability of Mark I_m and studying its performance both in a simulated environment and with robots, we make the sequencing problem harder and we introduce Mark II_m: Mark II_m assumes that a robot must perform a complete sequence before receiving any feedback on whether the sequence is correct or not. We measure the performance of both versions of the system with $m = 3$ (Mark I_3 and Mark II_3) and $m = 4$ (Mark I_4 and Mark II_4). The results that we provide show that TS-Swarm is indeed able to sequence tasks autonomously at operation time.

Because sequencing tasks is an albeit simple form of planning, TS-Swarm provides a new perspective on one of the most pivotal debates in the history of artificial intelligence: the debate on planning in robotics. This debate opposes two competing, antithetical paradigms: the deliberative and reactive (Murphy, 2000). According to the former, an intelligent robot should necessarily plan a course of action by reasoning on a model (Nilsson, 1984). According to the latter, a robot is more effective in dealing with the world by simply reacting to contingencies, without relying on reasoning and representation (Brooks, 1991). Although hybrid systems have been proposed, they conceptually juxtapose the two paradigms: deliberative and reactive instances—operating sequentially or in parallel—interact but remain logically distinct (Arkin, 1990; Saffiotti et al., 1995). By contrast, TS-Swarm associates the two paradigms in a novel and particular way: the ability to plan emerges at the collective level from the interaction of robots that, at the individual level, behave reactively without relying on reasoning and representation. TS-Swarm thus overturns traditional wisdom that robots can only either plan ahead using deductive reasoning or react to contingencies following pre-

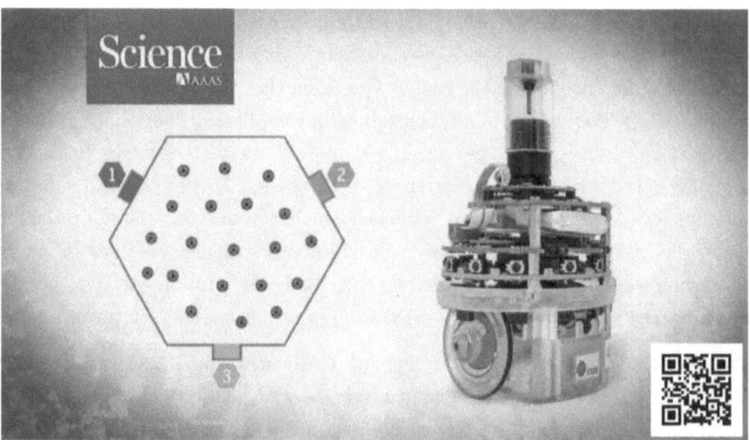

Figure 1.2: **Simple robots form a chain gang to solve complex problems.** Still from the video produced and published by Science. To watch the video, either click on the QR code in the bottom-right corner or scan it through the camera of a mobile device.

programmed rules. This message was disseminated also by Science, the world's leading outlet for scientific news, through a news article (Sutton, 2018) and a video about our work that was published on their YouTube channel. Fig. 1.2 shows a still extracted from the video and a QR code that links to it.

The rest of the **book** is organized as follows: In Chapter 2, we provide an overview of the domain of swarm robotics from an engineering perspective. We present the different approaches that researchers have adopted to address the challenges of designing robot swarms. We describe the most common collective behaviors that have been developed in the literature, mostly devoted to the emergence of geometrical/spatial properties and mechanical abilities. We discuss some notable systems that can be identified as milestones in the history of swarm robotics. Finally, we provide a discussion on cognition, focusing on both the way it has been pursued in artificial intelligence in general, and in swarm robotics more in particular.

In Chapter 3, we present the ideas at the basis of TS-Swarm and the implementation of the first version of the system: Mark I. We describe in details the different roles that the robots can assume at operation time, providing insight into how the system is ultimately able to autonomously sequence tasks. We then present our experimental setup and we analyze the results obtained by Mark I_3 and Mark I_4. Finally, we discuss some possible future improvements.

In Chapter 4, we make the task-sequencing problem harder and we present the second version of the system: Mark II. While in Mark I a robot receives a feedback as soon as it performs a task, in Mark II a robot must perform a complete sequence before receiving any feedback on whether the sequence is correct. We first discuss the implications of this change on the complexity of the problem: because of the lack of an immediate feedback, the problem faced by Mark II$_m$ is combinatorial and its computational complexity is $O(m!)$. Then, we describe the differences between the implementations of Mark I and Mark II. Finally, we present the experiments performed on Mark II, we analyze the results, we compare them with those of Mark I, and we draw some conclusions. Also in the case of Mark II, we discuss some possible improvements to the system.

In Chapter 5, we conclude the book. We provide a summary of the contributions and we discuss their relevance to the fields of swarm robotics and artificial intelligence. Finally, we suggest and discuss some future research directions.

Chapter 2

State of the art

This chapter presents the swarm robotics literature providing an overview of its most important contributions. Particular focus is on the problem of designing robot swarms that are capable of accomplishing a given mission in autonomy. This requires robot swarms to possess both practical abilities to solve the mission at hand, and cognitive abilities to autonomously make decisions based on the unpredictable circumstances encountered. An in-depth discussion on the problem of cognition in artificial intelligence, and in swarm robotics in particular, is developed in Section 2.5.

Additionally, the chapter describes a set of collective behaviors that have been demonstrated in the literature and few particularly notable robot swarms that serve as concrete examples of the recent achievements in the field of swarm robotics.

A complete review of swarm robotics from an engineering perspective can be found in Brambilla et al. (2013). Other articles have previously reviewed the swarm robotics literature: Şahin (2005) was the first to formally define the basic concepts of swarm robotics and to provide a survey of the literature. Bayindir and Şahin (2007) presented the literature via five taxonomies: modeling, behavior design, communication, analytic studies and problems. Iocchi et al. (2001) classified multi-robot systems depending on their degree of awareness, coordination, and decentralization and dedicates a section to applications of multi-robot systems. Gazi and Fidan (2007) surveyed the literature from a control-theory perspective, focusing on the problems of modeling the dynamics of a robot swarm, and presenting approaches for its control and coordination. Finally, Schranz et al. (2020) presented a complete survey of the latest collective behaviors developed and described a set of prospective applications for robot swarms.

This chapter of the **book** is organized as follows: Section 2.1 introduces the field of swarm robotics; Section 2.2 presents the most common approaches used to design collective behaviors for robot swarms; Section 2.3 describes a number of collective behaviors that have been realized and discussed in the literature; Section 2.4 describes six notable robot swarms that have been demonstrated. Section 2.5 discusses the problem of endowing robots with cognitive abilities and describes how the problem has been tackled so far by the swarm robotics community.

2.1 Swarm robotics

Swarm robotics is an approach to robotics in which a mission is entrusted to a large group of robots, a so called *robot swarm*. A robot swarm operates in an autonomous and self-organized way. A swarm does not rely on any centralized entity for making decisions and for coordinating its activities. In particular, a swarm does not rely on a leader robot or on external infrastructures: the collective behavior of the swarm is the result of the interactions between the individual robots and between robots and environment.

A characteristic of robot swarms is locality of interaction: each robot has a limited range of communication and perception. As a consequence, at any moment in time, each robot directly interacts only with the other (relatively few) robots that happen to be in its neighborhood.

In a typical swarm robotics application, robots operate in parallel on multiple tasks. They switch from task to task according to contingencies. Coherently with what has been stated above, task allocation is autonomous, self-organized, and based only on locally available information.

Autonomy, self-organization, redundancy, locality, and parallel execution are highly appreciated characteristics of a robot swarm as they are commonly deemed to promote fault tolerance, scalability, and flexibility.

Fault tolerance: high redundancy and the lack of a single point of failure (no leader robot, no external infrastructure) promote the realization of a system that is robust to failures of individual robots.

Scalability: locality of interaction promotes the realization of a system in which the addition (or removal) of robots does not qualitatively change the behavior of the system and therefore does not require modifying the behavior of the individual robots— provided that robot density is not dramatically altered.

Flexibility: parallel execution and autonomous task allocation promote the realization of a system that reacts and adapts to contingencies, modifications of the environment, and variations of the working conditions.

Swarm robotics juxtaposes itself to the single-robot approach (Nilsson, 1984) and to classical multi-robot approaches (Dudek et al., 1996; Parker, 2000; Iocchi et al., 2001).

In the single-robot approach, a mission to be performed is entrusted to a single, monolithic robot, for example, a humanoid robot. With respect to the classical single robot approach, swarm robotics appears to be more promising in applications in which fault tolerance, scalability, and flexibility are particularly desirable. Moreover, each of the individual robots composing a swarm is mechanically simpler than a single monolithic robot whose capabilities are comparable to the one of the swarm. As a consequence, it should be expected that the cost of hardware design is reduced in the case of swarm robotics.

On the other hand, designing the individual robot behavior that, through robot-robot and robot-environment interactions, would produce the desired collective behavior is more complex than designing the behavior of a single, monolithic robot. In the classical multi-robot approaches, the mission of interest is entrusted to a relatively small team of robots (Gerkey and Matarić, 2004; Dudek et al., 1996), smaller than a typical swarm. Usually, the team behaviors are tailored to the specific team size and thus need to be adjusted as the team size varies. In the classical multi-robot approaches, each team member has a role that is defined at design time. Also the patterns of interaction are defined at design time and are typically more rigid than those that characterize a robot swarm. As a consequence, a classical multi-robot system is not as fault tolerant, scalable, and flexible as a robot swarm. On the other hand, the interaction protocols of a classical multi-robot system are typically simpler to define than those of a robot swarm, as interactions are well defined and predictable and all relevant information is available at design time.

Beside being a promising engineering approach to the development of complex robotics systems, swarm robotics can be a powerful tool for studying social behaviors in biology as it is attested by a significant body of literature (Mitri et al., 2013). When swarm robotics is used as a tool to study social behaviors, the robots are programmed to reproduce as faithfully as possible the behavior observed in the biological system under analysis.

This **book** presents swarm robotics from an engineering perspective with the ultimate goal of adopting robot swarms in real-world applications.

When swarm robotics is intended as a field of engineering, social behaviors of insects, birds, and mammals are often a valuable source of inspiration for the designer. Nonetheless, as the goal of a designer is the pragmatic one of producing a system that accomplishes a given mission, the source of inspiration is loose and the designer is ready to depart from the biological system should this be needed to meet the requirements. In an engineering perspective, the biological plausibility of the final result is not a value in itself.

2.2   Design

Designing a robot swarm that is able to accomplish a given mission is a difficult endeavor. Usually, requirements on the mission are expressed at the swarm level, the so called *macroscopic* level, while the designer works at a lower level by implementing the behavior of the individual robots that compose the swarm, the so called *microscopic* level. The interaction of the individuals gives rise to a collective behavior that should satisfy the swarm-level requirements. In particular, the resulting collective behavior should allow the swarm to accomplish the given mission. To date, no general formal method exists to derive the individual behavior from the swarm-level requirements. The problem of designing robot swarms is tackled either manually or via automatic methods.

2.2.1   Manual design

In manual design, the designer of the swarm develops, by hand, the behavior of the individual robots that yields the desired collective behavior. In swarm robotics, the behavior of the individual robots is typically reactive—that is, robots act in response to contingencies (possibly influenced by their memory), without planning their future actions nor reasoning on their effects. The software architecture that is most broadly adopted is a particular class of finite state machines, the probabilistic finite state machine (Rabin, 1963).

Although most robot swarms are still developed through trial and error, in recent years principled design approaches have been proposed. The following two sections are devoted to the trial-and-error design approach and to some of the most promising principled design approaches.

Trial-and-error design

Designing a robot swarm by trial and error is more of an art than a science. The designer operates in an unstructured way with little scientific basis and technical tools: the designer searches for an individual-level behavior that, through the complex interaction of a large number of robots, would result in the desired collective behavior. The search process is performed via educated guesses that rely solely on the expertise and the ingenuity of the designer.

The designer starts by defining a first implementation of the individual robot behavior. The designer then tests the behavior, usually by means of computer-based simulations, and iteratively adjusts it until the resulting collective behavior meets the swarm-level requirements.

Often, the designer takes inspiration from biological systems: when the goal is to design a robot swarm whose swarm-level behavior is similar to the one of a biological system (e.g., a swarm of insects, a flock of birds, or a herd of mammals), the designer might find convenient to design the behavior of the individual robot by mimicking the one of the individual member of the biological system.

The relationship between the microscopic and the macroscopic levels poses challenging issues to the trial-and-error design approach. In particular, the behavior of the individual robot cannot be evaluated directly and *per se*: it must be evaluated indirectly by observing the collective behavior of a swarm composed by a large number of individual robots that execute the behavior under analysis.

Notwithstanding its limitations, the trial-and-error approach has been and still is the most successful approach to designing complex collective behaviors for robot swarms. Some examples of collective behaviors obtained by trial and error include aggregation (Şahin, 2005), chain formation (Nouyan et al., 2009), and task allocation (Liu and Winfield, 2010). These behaviors are described in more detail in Section 2.3.

Principled design

Although a general engineering framework for designing robot swarms is not available yet, a number of promising principled design methods have been proposed. These methods borrow concepts and tools from different disciplines and address different issues.

In virtual physics-based design (Spears et al., 2004) each robot is considered as a virtual particle that exerts forces on other particles—that is, other robots. Each robot is thus immersed in a field of forces that depends on the presence and distance of neighboring robots. The virtual force acting on each robot is $\mathbf{f} = \sum_{i=1}^{k} f_i(d_i)e^{j\theta_i}$, where j denotes the imaginary unit, d_i and θ_i are the distance and the direction of the ith neighboring robot, and the function $f_i(d_i)$ is the derivative of an artificial potential function. The Lennard-Jones potential (Jones, 1924) (Figure 2.1A) is commonly used in this context (e.g., Spears et al., 2004; Spears and Spears, 2012; Pinciroli et al., 2008). Figure 2.1B,C,D shows three examples of the virtual force that acts on a robot depending on the position of its neighboring peers. The designer can associate virtual repulsive forces to obstacles and other objects in the environment to prevent collisions.

Each robot estimates the virtual forces that act on it and translates them into motion commands. The main benefit of virtual physics-based design is that it allows the designer to formally prove properties of swarm-level behaviors including stability, convergence, and robustness. An extension of virtual physics-based design is the Hamiltonian method (Kazadi, 2009). Starting from a mathematical description of the

swarm at the macroscopic level, the Hamiltonian method derives the microscopic behavior that minimizes or maximizes the value of a relevant quantity (e.g., the virtual
potential energy of the state of the swarm). The major drawback of virtual physics-
based design and of the Hamiltonian method is that they are suitable only for designing
spatially-organizing collective behaviors (see Section 2.3.1).

Control theory is the theoretical framework of a few principled design methods
that have been proposed. Some of these methods combine virtual physics with sliding
mode control to design robot swarms that perform aggregation, foraging, and pattern
formation (Gazi, 2005; Gazi and Passino, 2004b) (see Section 2.3). Other methods
use kinematic equations to model the motion of robots and a set of control-Lyapunov
functions to develop an individual behavior for pattern formation (Ögren et al., 2001;
Egerstedt and Hu, 2001) (see Section 2.3). The main advantage of methods based on
control theory is that some properties of the resulting robot swarms (e.g., stability and
robustness) can be proved using theoretical tools such as Lyapunov stability theory.
However, the application of control theory typically relies on assumptions—such as
deterministic behavior, global communication, and full synchronization—that are often
unrealistic in swarm robotics.

Defining a general link between the desired swarm-level (macroscopic) behavior and
the individual (microscopic) behavior is the key issue in the principled design of robot
swarms. Some studies showed that, under a series of assumptions, a microscopic implementation can be derived from a macroscopic model (Berman et al., 2009, 2011).
Through analytical means, the parameters of a macroscopic model described by a set
of advection-diffusion-reaction partial differential equations are mapped onto the individual behavior.

This method was successfully used to design robot swarms that perform task allocation and area coverage. However, the underlying assumptions—such as infinite number
of robots and global communication—are often violated in swarm robotics. As a result,
the macroscopic model often fails in predicting the performance of the final system, as
shown by a comparative experimental analysis (Mermoud et al., 2014).

Another study proposed a method to design the individual switching probabilities
in task allocation under soft deadlines (Khaluf et al., 2014). The total amount of work
performed by the swarm is described as a Poisson process. Via formal means, the
proposed method derives off-line the switching probabilities—that is, the parameters
of a Markov chain that describes the individual behavior. The method is based on the
assumption that the size and the deadline of the target tasks are all known at design
time.

A formal method borrowed from supervisory control theory has been used to design

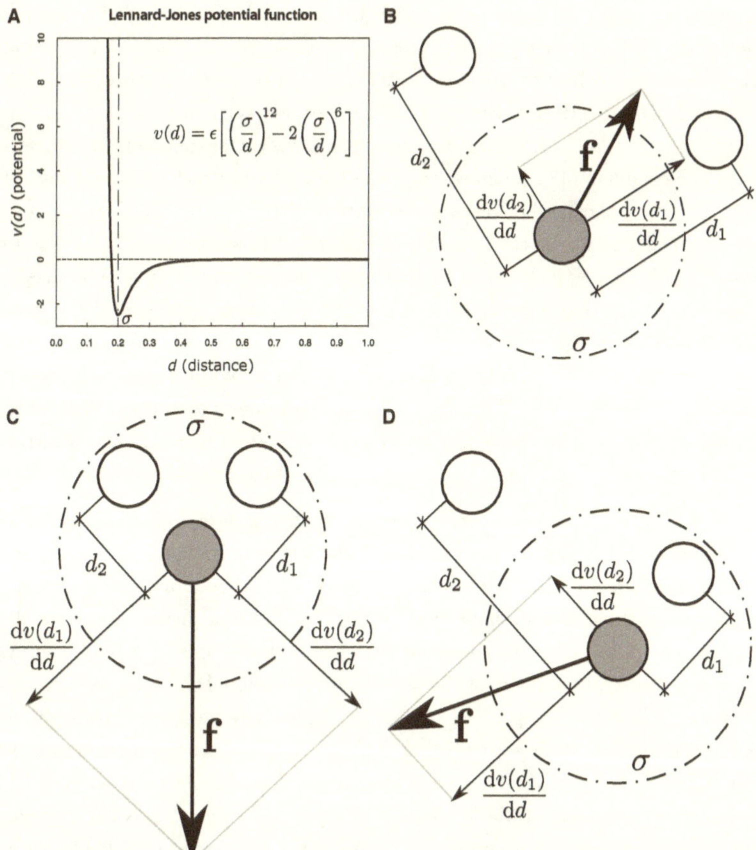

Figure 2.1: **Virtual physics-based design.** (**A**) The Lennard-Jones potential function. The potential v depends on the current distance d between two robots. σ is the desired distance between the robots. ϵ is a parameter called well depth and corresponds to the depth of the potential function. In this example, $\sigma = 0.2$ and $\epsilon = 2.5$. **B**, **C**, and **D** are examples of the virtual force that acts on a robot (gray-filled circle) depending on the position of two neighboring robots (white-filled circles). In **B**, the two neighboring robots are farther than the desired distance σ. Therefore, the robot is attracted by the two neighbors with forces that are determined by the derivative of the Lennard-Jones potential function at the point given by the distance of the neighbors. The resulting force is the sum of the individual forces. In **C**, the two neighboring robots are closer than the desired distance σ and hence exert repulsive forces on the robot. In **D**, one neighbor is at a distance $d_1 < \sigma$ and thus exert a repulsive force on the robot, while the other neighbor is at a distance $d_2 > \sigma$ and thus attracts the robot.

a segregation behavior (Lopes et al., 2014). The method is intended to be platform independent: the behavior produced was shown to successfully accomplish the mission both when executed by a swarm of e-pucks (Mondada et al., 2009) and when executed by a swarm of kilobots (Rubenstein et al., 2014a). Supervisory control theory is a method largely applied in manufacturing for automatically synthesizing control software that drives the behavior of a plant so that specifications are met. In its adaptation to swarm robotics, the method uses a description of the swarm-level requirements and a set of specifications for the behavior of individual robots to generate the individual control logic. In particular, the method requires that the designer specifies the set of possible events that may occur and the corresponding responses of the individual robots necessary to engender the desired collective behavior. In other terms, the rules that trigger the transition from task to task in the individual robot's behavior must be known at design time. Therefore, this method does not support the designer in the most crucial step: devising the appropriate individual behavior that generates the desired swarm-level behavior.

Another principled design method is property-driven design (Brambilla et al., 2014). Property-driven design was introduced with the ultimate goal of deriving an individual behavior from swarm-level requirements. The method is based on prescriptive modeling and model checking. The design process is composed of four phases: First, the designer defines a set of desired properties that the swarm should meet. Second, the designer produces a prescriptive model of the swarm and uses model checking to verify that the model complies with the specified properties. Third, the designer implements a simulated version of the robot swarm using the prescriptive model as a blueprint. Fourth, the designer implements the final robot swarm and validates the previous steps. Models are described by means of Markov chains and properties are defined by statements in probabilistic computation tree logic, a probabilistic temporal logic that captures well both temporal and stochastic aspects. Property-driven design is a structured design process supported by formal tools. However, the step from the prescriptive macroscopic model and the correspondent microscopic implementation is not yet automatic, and it is still reliant on the intuition of the designer.

Rather than defining a unifying framework to obtain any possible collective behavior, a number of works have proposed the idea of defining a catalogue of design patterns (Babaoglu et al., 2006; Gardelli et al., 2007; Reina et al., 2015a). In the context of swarm robotics, a design pattern is a collection of guidelines to obtaining a specific collective behavior. It must provide: i) a macroscopic model that describes the swarm-level requirements, ii) a description of the microscopic behavior, and iii) a mapping from the parameters of the macroscopic model to those of the microscopic behavior. A first example of a design pattern for collective decision-making has been proposed and successfully used to design a collective foraging behavior (Reina et al., 2015b).

An interesting design method has been proposed to obtain self-assembly (Rubenstein et al., 2014b) (see also Section 2.3) and construction (Werfel et al., 2014) (see also Section 2.3 and Section 2.4). The method promotes the decomposition of the design problem (Nagpal, 2002): First, the user provides a global description of the desired aggregate/structure (e.g., the shape of the aggregate, the height of the structure). Second, the method defines a set of steps necessary to build the desired aggregate/structure. Third, the method maps these steps onto individual rules (e.g., rules of motion along the border of the aggregate or over the structure). The method enables the validation of some properties of the final system, such as correctness and convergence. It has been applied successfully to self-assembly and construction. Unfortunately, its applicability is limited to a restricted set of missions.

Finally, although programming and scripting languages cannot be considered as design methods on their own, they can significantly ease the principled design of robot swarms. Two prominent examples are Protoswarm (Bachrach et al., 2010) and Buzz (Pinciroli et al., 2015). Protoswarm (Bachrach et al., 2010) is a scripting language based on the abstraction of an amorphous computational medium (Beal, 2004). The amorphous computational medium assumes that the environment is filled with entities that can compute and communicate locally with each other. Protoswarm enables the definition of behaviors for the individual robots by writing scripts at the level of the swarm. The scripting language features swarm-level primitives that deal both with space and time. These primitives are translated approximately into individual robot behaviors by a runtime library. The idea of languages based on manipulations of computational mediums has received increasing attentions from the research community.. New languages have been proposed for the creation of swarm-level programs with a sound mapping between the swarm-level and the individual-level primitives (Viroli et al., 2013; Beal and Viroli, 2015).

On the other hand, Buzz (Pinciroli et al., 2015) is a programming language for heterogeneous robot swarms. Buzz offers primitives to work either at the individual level or at the swarm level. For the time being, the swarm-level primitives mostly serve to create different teams and assign robots to each team. Moreover, Buzz provides mechanisms to share information locally and globally, thanks to a virtual stigmergy mechanism based on a distributed tuple space. One of the prominent features of Buzz is its modularity: primitives can be combined or defined anew to create modules that can be tested, compared, and reused.

Finally, some work has been devoted to proposing new constructs for well-established modeling languages. In Cavalcanti et al. (2018), the authors propose an extension of RoboChart (a UML-based language for single-robot systems) for modelling and verifying swarms of interacting robots.

2.2.2   Automatic design

In automatic design methods, the individual behavior of the robots that compose the swarm is generated automatically through an optimization process. The burden of searching for the individual behavior that results in the desired collective behavior moves from the designer to a computer program. The majority of work on automatic design of control software for robot swarms has been produced within the evolutionary robotics domain. A few alternative methods have been proposed.

Evolutionary robotics

Evolutionary robotics (Nolfi and Floreano, 2000) takes inspiration from the Darwinian principles of natural selection and evolution to automatically design control software for single and multi-robot systems. In evolutionary robotics, the design process starts typically from a population of behaviors generated at random. At each iteration, the behaviors are evaluated over a set of experiments, typically via computer-based simulations. The same behavior is used as control software for all the robots of the swarm. The evaluation is performed by a fitness function that measures the performance of the swarm. The best scoring behaviors are used to produce the next generation by means of genetic operators: cross-over, which mixes treats from parent solutions to produce a child behavior from them, and/or mutation, which alters small treats of single behaviors. The process terminates when a time limit or a certain performance threshold are attained or when the fitness function stops improving.

The individual behavior can be represented in several ways, but the most common one is via an artificial neural network. The evolutionary process searches the parameters of a neural network which, when used as control software on all the robots, maximizes the performance of the swarm.

Despite the large number of works that showed the effectiveness of evolutionary techniques, an engineering methodology for the application of evolutionary robotics is still unavailable (Trianni and Nolfi, 2011; Silva et al., 2016). The main issues are that the evolutionary process does not give any guarantee of convergence and the neural networks that result from the design process are black-boxes: they are difficult to analyze and understand. Moreover, the challenge for the designer is often simply shifted: from devising the desired individual behavior to devising the right fitness function which would guide the optimization algorithm towards the desired individual behavior. For these reasons, most of the behaviors produced so far via evolutionary robotics are relatively simple and thus easily obtainable via manual design. Some promising ideas have been proposed that could contribute to the development of an engineering methodology for evolutionary robotics. Multi-objectivization is deemed to improve the effectiveness of the design process by guiding the evolutionary search in rugged fitness landscapes (Trianni and López-Ibáñez, 2015). Novelty search (Lehman and Stanley, 2011) is deemed to promote diversity among candidate behaviors and improve the exploration of the search space (Gomes et al., 2013). Finally, the hierarchical decomposition of the control software into modules is deemed to ease the design process (Duarte et al., 2014a,b). Some efforts have also been devoted to evolving collective behaviors directly on the robots (Jones et al., 2019), rather than in simulation. In this work, a distributed evolutionary algorithm (island model distributed evolutionary algorithm) running on the physical swarm was used to automatically design a foraging behavior. Besides the onboard evolution, this work also attempted to make the final behaviors human-understandable by using behavior trees as the controller architecture. For a comprehensive review and critical discussion of the evolutionary robotics literature, see (Trianni, 2014; Silva et al., 2016).

Reinforcement learning

In reinforcement learning, agents try to select the course of action that maximizes a cumulative reward through repeated interactions with their environment. At each interaction, agents may receive a positive or negative feedback (reward) for their actions. Reinforcement learning is widely adopted in robotics. It has been elegantly defined and successfully used in single-robot scenarios (Kaelbling et al., 1996; Sutton and Barto, 1998). The multi-robot case has been considered only by few works with limited scope (Panait and Luke, 2005; Bušoniu et al., 2010). Swarm robotics appears to pose major problems to reinforcement learning and only a very limited number of studies have been proposed (Matarić, 1997, 1998). The main problem is the decomposition of the collective reward into the rewards that should be assigned to the individual robots: How does a particular action performed by an individual affect the collective behavior of a complex system such as a robot swarm? Moreover, the complexity that arises from the great number of interactions between robots in a swarm leads to the explosion of the state space size. For these reasons, the results are limited to specific tasks and have been demonstrated in experiments with only few robots.

A form of reinforcement learning that seems more promising for facing the problems posed by the design of robot swarms is team learning (Panait and Luke, 2005; Bușoniu et al., 2008). In team learning, the learning process takes place at the collective level rather than at the individual one. As the learning process is a single one, team learning can use standard single-agent machine learning techniques. However, the major problem remains the explosion of the state space size due to the large number of interactions between individuals.

More recent approaches to reinforcement learning use deep neural networks and explore techniques of transfer learning between robots and between tasks to mitigate the problems discussed (Devin et al., 2017). The preliminary results in simulation are promising but more extensive tests with real robots and swarms of larger size are required.

Other methods

Because of the limitations of evolutionary robotics (Trianni and Nolfi, 2011; Trianni, 2014), other automatic design methods for robot swarms have been proposed in the recent years.

A number of studies focused on on-line adaptation in multi-robot systems. In these studies, the execution of population-based algorithms is distributed over a group of robots (Watson et al., 2002). In this form of embodied evolution the robots are used as computation nodes. Several works have tested the feasibility of this approach, proposing different solutions, including open-ended and task-dependent evolution and the use of finite state machines (Bredeche et al., 2012; Haasdijk et al., 2014; König and Mostaghim, 2009). The implementation of distributed evolutionary algorithms in robot swarms has been tested in other variants: some study explored the idea of cultural evolution in robot swarms using an imitation-based algorithm (Winfield and Erbas, 2011). The particle swarm optimization algorithm was compared to genetic algorithms for on-line adaptation and proven to provide a higher degree of diversity in the robot swarm (Pugh and Martinoli, 2007; Di Mario and Martinoli, 2014).

Another promising and effective approach that has been proposed adopts a fixed control architecture and focuses on tuning only a small set of parameters. Genetic algorithms and evolutionary strategies were used to optimize the parameters of finite state machines for a cooperative foraging and object clustering (Hecker et al., 2012; Gauci et al., 2014a). Exhaustive search was used to determine the optimal parameters for self-organized aggregation (Gauci et al., 2014b). A similar approach combines evolutionary computation with virtual-physics based design to learn off-line the parameters for the Lennard-Jones potential function in a navigation task (Hettiarachchi, 2007).

Another promising approach to the automatic design of control software for robot swarms is AutoMoDe (Francesca et al., 2014). In AutoMoDe, the control software is automatically designed in the form of a probabilistic finite state machine. The design process works by combining and fine-tuning preexisting modules, which have parameters that regulate their functioning. A search algorithm optimizes these parameters along with the topology of the probabilistic finite state machines to maximize a task-dependent performance measure. This design method was proven effective in overcoming the reality-gap and in subsequent studies was shown to outperform human designers in designing control software for five different missions (Francesca et al., 2015).

For a complete discussion and review of the achievements in the automatic design of robot swarm we refer to Francesca and Birattari (2016) and Bredeche et al. (2018). A manifesto of the current status of research and future challenges can be found in Birattari et al. (2019). Finally, Birattari et al. (2020) divided optimization-based design methods into two categories: semi-automatic design, in which the human designer actively guides the design process; and fully-automatic design, in which the design process does not involve any human intervention.

2.3 Collective behaviors

Collective behaviors are basic behavioral units of robot swarms that can be combined to obtain robot swarms that are able to perform complex missions. In particular, here we describe the collective behaviors that display geometrical/spatial properties and mechanical abilities. These are the behaviors on which most of the research has focused so far. We divide these collective behaviors into three categories: spatially-organizing behaviors, navigation behaviors, and interaction with humans. Another section (Section 2.5.2) is entirely devoted to the description of collective behaviors that display simple cognitive abilities.

2.3.1 Spatially-organizing behaviors

Spatially-organizing behaviors are collective behaviors that focus on how the robots distribute and organize in space.

Aggregation

The goal of aggregation is to group the robots in a region of the environment. Aggregation is a useful building block for many complex behaviors as it allows robots to gather and thus to interact with each other. The implementation of aggregation in robot swarms is often inspired by similar behaviors observed in natural systems such as bacteria, bees, and cockroaches. Aggregation has been obtained with either manual or automatic design methods.

Manual design methods typically adopt a simple probabilistic finite state machine: the robots wander in the environment and, when they find other robots, they decide stochastically whether to stay in their proximity or depart from them. Typically, robots join an aggregate (or leave it) with a probability that is a function of the size of the aggregate itself: the larger the aggregate, the higher the probability of staying. This favors the formation of a single, large aggregate, as small aggregates tend to disband. This basic behavior can be adapted and tuned to obtain either static or moving aggregates (Garnier et al., 2005; Soysal and Şahin, 2005). A later study (Firat et al., 2020) showed how these aggregation mechanisms can be guided by introducing in the swarm a small portion of informed robots—that is, robots that stop only on the designer pre-defined site/s for aggregation. Aggregation has been obtained also via a principled design method based on control theory (Gazi, 2005) (see Section 2.2). Automatic design methods mostly use artificial evolution to find the parameters of a neural network that produces the desired aggregation behavior. Either static or moving aggregates can be obtained with this approach (Trianni et al., 2003; Soysal et al., 2007). Other automatic design approaches work on a fixed control architecture and tune a small set of parameters. This approach successfully produced an aggregation behavior with memoryless robots that are equipped only with a single binary sensor (Gauci et al., 2014b).

Aggregation can be modeled using different modeling techniques. Rate equations are particularly suited because of their ability to describe the evolution in time of the portion of robots in a particular state (the aggregate) (O'Grady et al., 2009b). Other modeling methods used in the literature are based on Langevin and Fokker-Planck equations (Hamann and Wörn, 2008; Schmickl et al., 2009), on Markov chains (Soysal et al., 2007; Correll and Martinoli, 2011), and on control and stability theory (Gazi and Passino, 2003, 2004a).

Pattern formation

Pattern formation is a behavior that aims at positioning robots in space according to a certain, well defined, pattern. Pattern formation can be useful for a number of purposes such as covering an area, achieving a certain network topology and forming the initial configuration for coordinated motion (see Section 2.3.2). Examples of pattern formation that often inspire research in swarm robotics can be found both in biology (e.g., the chromatic patterns on some animal's coat) and physics (e.g., crystal formation and Bénard cells).

Pattern formation in robot swarms is typically obtained using virtual physics-based design. As already mentioned in Section 2.2, in virtual physics-based design robots are considered immersed in the virtual potential field generated by the neighboring robots. Motion commands are computed by each robot based on the sum of the virtual forces exerted by its neighbors.

If all the robots exert the same force, this simple mechanism yields an hexagonal lattice (Spears et al., 2004). By dividing the swarm in two groups with different attraction/repulsion thresholds, it is possible to obtain a square lattice (Spears and Spears, 2012; Pinciroli et al., 2008). Virtual physics can be combined with tools borrowed from control theory. In this case, the stability of the resulting formation can be proved analytically (Gazi, 2005; Egerstedt and Hu, 2001) (see Section 2.2).

Virtual springs can be used alternatively to compute the forces of attraction and repulsion. Combined with different interaction rules (e.g., full connectivity, nearest neighbor, K-nearest neighbors), they can produce different patterns (Shucker and Bennett, 2007; Shucker et al., 2008).

Notably, a pattern formation behavior with a thousand robots has been demonstrated (Rubenstein et al., 2014b). Few robots act as the seed of the pattern and define the origin and orientation of the coordinate system that is used to build the desired shape. Starting from the seed robots and using also an internal representation of the target pattern, other robots of the swarm gradually join the pattern. Robots localize themselves with respect to the initial seed using an information gradient. The thousand robots have been shown to successfully form different shapes.

Another work showed how it is possible to define a simple local behavior that enables the swarm to eventually arrange in a global desired pattern (Coppola et al., 2019). The behavior was designed with the twofold goal of guaranteeing convergence to the desired patter while assuring safety (collision avoidance and coherence in the swarm).

An important application of pattern formation is area coverage: when the number of robots is limited, a lattice formation of equally-spaced robots optimizes the coverage of the space (Howard et al., 2002). Area coverage is often modeled using differential equations: two examples of differential equations used to model area coverage are a set of advection–diffusion–reaction partial differential equations (Berman et al., 2011) and the Fokker–Planck equations (Prorok et al., 2011). In the latter example, the accuracy of four models based on Fokker-Planck equations was tested by comparing their predictions with the results of computer-based simulations and real-robot experiments. A solution for area coverage applicable to extremely simple robots (e.g., robots that lack computational power and/or storage) was recently demonstrated with a swarm of 25 e-pucks (Özdemir et al., 2019).

Chain formation

In chain formation, robots arrange themselves in the environment to create a chain
that connects two locations. The chain is then used by other robots as a navigation
aid (see Section 2.3.2). This behavior is inspired by Argentine ants, which form chains
of individuals that connect their nest to foraging sites (Deneubourg et al., 1990).

Chain formation can be developed using different design methods. Typically it is
obtained by manually designing control software in the form of a probabilistic finite
state machine. The chain is built incrementally from the starting location. The robots
that find a growing chain follow it until the end and join it in the last position with
a certain probability. The last robot in the chain can always leave the chain with a
certain probability. This prevents the chain from becoming entrapped in dead ends
and allows an effective exploration of the environment. When the chain reaches the
target location, it becomes stable. The robots in the chain might use a tricolor-pattern
to indicate the direction of the chain (Nouyan et al., 2008, 2009). A variant of this
solution is based on a probabilistic finite state machine and network routing. The
result is a chain of moving robots (Ducatelle et al., 2011b).

Virtual physics-based design and automatic design methods can also be used to
design chain formation. In virtual physics, virtual forces are used to maintain a desired
distance between robots in the chain and between robots and walls in order to create
chains that strongly depends on the shape of the environment (Maxim et al., 2009).
Concerning automatic design, artificial evolution has been shown able to produce chains
of moving robots (Sperati et al., 2010).

Self-assembly and morphogenesis

Self-assembly is the process in which robots physically connect to each other. Self-
assembly can be useful, for example, to increase mechanical stability and ease navigation
on rough terrains. When the connected robots form a pattern or shape, the process
is called morphogenesis. Morphogenesis is used when a particular structure allows the
swarm to perform a specific task. For instance, a line of connected robots can navigate
over a hole whereas a single robot would fall into it. Several natural systems show
self-assembly and morphogenesis behaviors: ants are able to create bridges, rafts and
walls to perform specific tasks; cells self-organize structures to form tissues and organs.

Self-assembly and morphogenesis can be designed in several ways. These behaviors
pose many challenges to the design process: When and how should the assembly process
start? Which robots should connect to each other? Which shape should be formed?
Each of these challenges can be addressed in different ways.

Robots can trigger the self-assembly process when they encounter obstacles or adversities that they are not able to overcome on their own. Empirical studies showed that connected robots are able to navigate in hazardous terrains better than individual robots (O'Grady et al., 2010), they can overcome obstacles that a single robot cannot overcome (Mondada et al., 2005), and they can transport heavy objects faster and for longer distances (Groß and Dorigo, 2009). Robot swarms have also been demonstrate capable of creating 3D structures through self-assembly (Levi and Kernbach, 2010).

Homogeneous robots can self-organize the assembly process by signaling the docking points in different locations of their bodies. Other robots can then connect stochastically to those docking points. In this way the robots can form different structures, such as lines, stars and circles (O'Grady et al., 2009a). Alternatively, the capabilities of heterogeneous robots can ease the process of self-assembly. For example, aerial robots can recognize the task to perform and indicate to the ground robots which robots should self-assemble and what structure they should create to perform the task (Mathews et al., 2012, 2019).

Self-assembly and morphogenesis have not been modeled often in the literature. A study showed that a self-assembly behavior that allows robots to form lines can be modeled using a set of chemical reactions (Evans et al., 2010). This set of chemical reactions was then abstracted by a set of differential equations, solved approximately by means of stochastic simulations (e.g., Gillespie algorithm), and compared to computer-based simulations. Other work uses bottom-up approaches to morphogenesis inspired by observed patterns in biological systems during embryogenesis (Slavkov et al., 2018). On top of this, subsequent work focused on enhancing the controllability of morphogenesis and showed how to produce richer shapes (Carrillo-Zapata et al., 2019).

Object clustering and assembling

Object clustering and assembling refer to behaviors in which the robots create aggregates of objects. The difference between object clustering and assembling is that in the former the aggregates are clusters of unconnected objects, whereas in the latter the objects must be connected by some kind of physical link. These two behaviors are at the basis of any swarm construction system. For the design of object clustering and assembling, researchers often take inspiration from social insects: brood clustering has been observed in ants, termites can build mounds that are orders of magnitude larger than the single individuals.

Object clustering is usually obtained using a probabilistic finite state machine. The robots explore randomly the environment and react with appropriate responses when they find an object or partially formed clusters. In the simplest form of object clustering, a robot picks up an object and deposit it with a probability that is proportional to the number of other objects perceived (Beckers et al., 2000). The final position of the clusters can be controlled by marking the ground with colors or using other signals recognizable by the robots (Melhuish et al., 1999b; Stewart and Russell, 2006). Object clustering can also be obtained via automatic design methods. A clustering behavior for extremely simple robots was successfully developed through evolutionary robotics (Gauci et al., 2014a): the robots are not capable of arithmetic computation and are only able to detect the presence of an object or another robot in their direct line of sight. Despite these limitations, the swarm is able to successfully create clusters of objects within a limited amount of time. Object clustering was modeled in a seminal work on the use of rate equations in swarm robotics (Martinoli et al., 1999).

Concerning assembling, a prominent work demonstrated a behavior that enables the creation of arbitrary 3D structures (Werfel et al., 2014). This solution generates off-line a set of traffic rules and assigns them to the robots, together with a static representation of the target structure. Respecting the traffic rules, a group of climbing robots builds the structure by placing a building block at a time. More details on this work can be found in Section 2.4.

2.3.2 Navigation behaviors

Navigation behaviors are collective behaviors that aim at coordinating the movements of a robot swarm.

Collective exploration

Collective exploration includes behaviors whose goal is to explore an environment, or interesting portions of it. Work on collective exploration takes frequently inspiration from behaviors observed in natural systems. Control software for collective exploration is typically implemented in the form of probabilistic finite state machines. Often the swarm relies on static robots that act as waypoints to guide the navigation of moving robots. To do that, the static robots can form either physical or virtual structures.

Physical structures are usually the result of pattern formation and chain formation (see Section 2.3.1). Once the physical structure is formed, the moving robots can follow it, waypoint after waypoint, to navigate in the environment. In virtual structures, the static robots are not necessarily close to each other, but they are connected by a virtual medium. For example, pre-deployed robots can create a virtual structure between two locations by exchanging messages. Moving robots can exploit these messages for navigation (Payton et al., 2001; Di Caro et al., 2009). Similarly, a network of pre-deployed sensors can be used by the robots to navigate towards their goal location (O'Hara and Balch, 2007). The navigation route is calculated by the robots via a distributed variant of the Bellman-Ford algorithm. An hybrid solution was developed in the Swarmanoid project (Stirling and Floreano, 2010) (see also Section 2.4). In this solution, a set of aerial robots deploy sequentially to form a chain, using the position of the previously deployed robots to determine their target position. Once deployed, the robots establish also a virtual structure by acting as communication relays.

Lastly, a solution has been proposed in which the robots of a swarm both navigate and guide the navigation of others, simultaneously (Ducatelle et al., 2014). While moving, the robots share navigation information between them and hence cooperatively guide each other towards a target location. The advantage of this solution is that it does not bind any robots to a specific location. All the robots can thus move and be involved in other tasks, possibly unrelated to navigation.

Coordinated motion

In coordinated motion, also known as *flocking*, the robots move in formation through the environment, similarly to flocks of birds or schools of fish.

In nature, coordinated motion is used by many animals to reduce energy consumption and increase the chance they survive attacks of predators. Flocking can be obtained with either manual or automatic design methods. The most common design method uses virtual physics. Virtual forces of attraction and repulsion maintain a desired constant distance between the robots and a uniform alignment during the motion (Reynolds, 1987). The robots are capable of coordinated motion even in absence of a common goal, thanks to the sole knowledge of heading and distance of their neighbors (Turgut et al., 2008a). Under this configuration, it is sufficient to insert few "informed" robots to direct the movement of the other "uninformed" robots, and hence of the whole swarm, toward a goal (Çelikkanat and Şahin, 2010). Further works showed that this behavior does not require an explicit alignment rule, and thus robots do not need to perceive the orientation of their neighbors. The swarm is still able to navigate with and without the presence of informed robots (Ferrante et al., 2012). Flocking of a swarm of aerial robots was obtained through evolutionary robotics (Hauert et al., 2009). Without relying on any external infrastructure, the aerial robots establish and maintain a wireless communication network to connect a base station and a user station that are located on the ground.

In the literature, flocking is typically modeled using differential equations. An example is the application of a method based on a Fokker-Planck equation (Hamann and Wörn, 2008). In another study, researchers performed preliminary steps towards linking the models of flocking produced in statistical physics with the studies produced in swarm robotics (Turgut et al., 2008b). The authors focused on the alignment of robots and verified the existence of a phase transition between order and disorder that depends on the level of noise and on the neighborhood size. The results were validated using computer-based simulations.

Collective transport

Collective transport refers to a set of behaviors in which the goal of the swarm is to cooperatively move objects from one location to another. The objects are too heavy for a single robot, thus cooperation is necessary. Collective transport can be observed in ant colonies. To achieve collective transport, ants use a trial-and-error process in order to determine the right pulling/pushing direction (Kube and Bonabeau, 2000).

Collective transport is usually designed via manual methods or artificial evolution. Different strategies can be employed for transporting the object: robots can connect directly to the object and move it, they can connect to each other and then to the object, or they can surround the object and push it with their movement (Groß and Dorigo, 2009; Wilson et al., 2014). Consensus on the direction of movement and cooperation are achieved either through direct or indirect communication. For example, when direct communication is used, robots can agree on a common direction of movement by averaging their individual desired directions (Campo et al., 2006). When communication is indirect, robots can position themselves around the object depending on the position already taken by other robots (Wilson et al., 2014), or depending on an estimation of the forces applied by other robots on the object or on their own chassis (Baldassarre et al., 2006).

As an alternative to reaching consensus on the direction on movement, some robots can form a chain to connect the source and the destination of the objects (see Section 2.3.1). The chain is then used as navigation aid by other robots that transport the objects (Nouyan et al., 2006).

2.3.3 Interaction with humans

Robot swarms are designed to work autonomously and to act in a distributed way. These characteristics limit the degree of control that a human operator can exercise on the system. However, there are several cases in which forms of human control over the swarm could be necessary. Human-swarm interaction studies how a human operator can control a robot swarm and receive feedback from it. Studies in this field can be categorized on the basis of the nature of interactions that they propose.

The most common approach relies on an intermediate modeling layer between the operator and the swarm. Usually, the modeling layer produces an abstract representation of the robots and their environment that is then displayed to the operator through a graphical user interface. By acting on the GUI, the operator can select robots and send them commands. The selection can contain single robots (Bashyal and Venayag-amoorthy, 2008), or a group of robots, which can be selected, for instance, by drawing a rectangular zone that contains the robots in the GUI (Kolling et al., 2013). A robot controlled by a human operator is perceived by the swarm as just another robot, and thus the influence of the operator is very limited. This problem can be solved partially by a hierarchical communication architecture in which the operator sends orders to the selected robot, which is called "the sergeant" (Bruemmer et al., 2001).

Other studies focus on the use of augmented reality. Part of the studies that use augmented reality propose solutions only for the visualization of feedback from the robots to the operator. For instance, an optical see-through head-worn device receives robots' messages, analyses them and augments the environment with their represen-tation (Daily et al., 2003). Similarly, firefighters are helped in their mission by a robot swarm, which gives them direction information displayed by augmented hel-mets (Naghsh et al., 2008). Other works provide bi-directional communication solu-tions: through a device that display the augmented environment, the operator can also give commands to the robots by acting on the real-time video stream (Ghiringhelli et al., 2014).

Finally, there are studies that aim at realizing a direct interaction, without relying on intermediate modeling levels. In fact, creating and maintaining an updated model of the robots and their environment is a demanding task. It often requires ad-hoc infrastructures and it becomes intractable in dynamic (real) environments or when the number of robots is greater than few units. For these reasons, techniques of direct interaction based on gestures recognition, face engagement and speech recognition have been proposed. Combinations of such techniques are possible (Couture-Beil et al., 2010; Pourmehr et al., 2013; Nagi et al., 2014). Performing gesture recognition directly on the robots might lead to mismatches, and thus requires distributed consensus algorithms in order to reach an agreement of all the robots on the same gesture (Giusti et al., 2012). Other works proposed the use of external sensors, such as the Microsoft Kinetic sensor to give commands through gestures to a robot swarm (Podevijn et al., 2013).

2.4 Notable robot swarms

Within the vast production of fundamental research work in swarm robotics, a few systems have been shown to be able to perform complex missions. In the following, we describe a selection of six notable systems.

Swarm-bot (Nouyan et al., 2009) is a robot swarm composed of relatively simple robots, the s-bots, that are able to attach to each other—see Figure 2.2A. The ability to self-assemble, along with control algorithms inspired by self-organized behaviors of social insects, allow the swarm-bot to effectively adapt to its environment. For example, by self-assembling in different shapes, the swarm-bot can navigate through rough terrains and drag objects that are too heavy for single s-bots. The swarm-bot has been shown to be able to find a target, heavy object and retrieve it. In the first phase, the s-bots form a chain between the object and the nest (see chain formation in Section 2.3). In the second phase, a group of s-bots surround the object, self-assemble, and drag the object to the nest along the path described by the chain.

Swarmanoid (Dorigo et al., 2013) is a heterogeneous swarm composed of three types of robots: eye-bots, hand-bots, and foot-bots—see Figure 2.2B. Eye-bots are flying robots specialized in sensing the environment and providing an overview to foot-bots and hand-bots. Hand-bots can climb walls or other vertical surfaces and grab objects, but they cannot move on the ground without the help of other robots. Foot-bots are specialized in moving on the ground and transporting either objects or other robots. The Swarmanoid has been shown to be able to explore an unknown indoor environment, locate a target object (a book), and retrieve it. First, the eye-bots explore the environment, find the book on a shelf, and highlight the path to it. Then, the foot-bots transport a hand-bot to the shelf following the path indicated by the eye-bots. At this point, the hand-bot climbs the shelf, grabs the object, and returns to the ground where the foot-bots transport it back to the initial location.

TERMES (Werfel et al., 2014) is a robot swarm inspired by how termites construct mounds—see Figure 2.2C. TERMES allows a user to specify a high level representation of the target 3D structure. From the specified structure, an off-line software generates a set of traffic rules that direct the flow of robots over the growing structure and regulate the building activity. Essentially, this set of rules, called *structpath*, is a 2D representation of the structure in which each stack is annotated with its height and a travel direction between each adjacent pair of stacks. Thanks to the structpath and to a static internal representation of the target structure, a group of custom-designed climbing robots proceeds autonomously to the construction process by depositing one brick at a time. The effectiveness of the system has been shown both in simulation and with a real-world implementation.

Thousand-robot Swarm (Rubenstein et al., 2014b) is a robot swarm composed of 1 024 Kilobots—see Figure 2.2D. The Kilobot is a small, low-cost robot equipped only with vibration motors for movement and an infrared transceiver for communication and distance sensing. A swarm of 1 024 Kilobots was demonstrated capable of forming different user-specified patterns. The pattern is built gradually, starting from few central robots that act as a seed. The other robots travel along the edge of the pattern under formation and stop in a proper position, according to an internal representation of the target pattern. The thousand-robot swarm is the largest robot swarm demonstrated so far.

CoCoRo (Schmickl et al., 2011) is a heterogeneous swarm of underwater robots—see Figure 2.2E. The swarm is composed of a base station and two types of robots: Jeff robots and Lily robots. Jeff robots are fast searching robots, Lily robots are slow information carriers. A CoCoRo can monitor, search, maintain, explore and harvest resources in underwater habitats. The CoCoRo was shown able to locate an object and guide the base station to its position. The Jeff robots search the seabed, until one of them finds the target object and starts recruiting more Jeff robots. At this point the Lily robots start to build a relay chain between the base station and the cluster of Jeff robots. The base station can then navigate to the object location using the information carried by the chain.

BioMachines Lab's Aquatic Robot Swarm (Duarte et al., 2015; Christensen et al., 2015; Duarte et al., 2014c) is a swarm of 10 aquatic surface robots—see Figure 2.2F. The robots are equipped with few sensors and actuators and thus are relatively inexpensive. The control software was designed automatically using evolutionary robotics (see Section 2.2). Four collective behaviors were developed to perform four different tasks: flocking, clustering, dispersion, and area coverage. By combining these collective behaviors, the swarm was shown able to perform a complex mission of environmental monitoring: the robot swarm collectively navigates towards an area of interest, optimizes the coverage of the area, monitors the water temperature in the area, clusters, and finally heads back to the base.

2.5 Cognition

The robot swarms that we have described so far in this chapter share two common characteristics: i) at the collective level, they display spatial/geometrical properties or mechanical abilities, and ii) the tasks that the individuals must perform to obtain such collective abilities and the order in which they must be performed were known at design time. The designers could thus devise and implement, either by hand or by means of automatic techniques, the individual rules that would result in the desired collective abilities.

Hard coding the tasks and their order of execution in the robot's control software limits the autonomy and the flexibility of the resulting robot swarm: the real world is dynamic and unpredictable by nature, therefore it is impossible to foresee at design

Figure 2.2: **Notable robot swarms.** (**A**) **Swarm-bot** – Previously unreleased photo. (**B**) **Swarmanoid** – Still from the video: *Swarmanoid, movie.* (**C**) **TERMES** – Still from the video: *Designing collective behavior in a termite-inspired robotic construction team.* (**D**) **Thousand-robot Swarm** – Still from the video: *Programmable self-assembly in a thousand-robot swarm.* (**E**) **CoCoRo** – Still from the video: *TYOC#52/52: Final Demonstrator.* (**F**) **BioMachines Lab's Aquatic Robot Swarm** – Still from the video" *A Sea of Robots.* Each picture contains a QR code in the bottom-right corner. To watch a video, either click on the corresponding QR code, or scan it through the camera of a mobile device.

time all the contingencies that the robots will face at operation time and the required countermeasures. The idea that we propose in this **book** is that robot swarms should be endowed with cognitive abilities, which allow them to modify and adapt their behavior at operation time depending on the contingencies encountered.

The problem of endowing artificial systems with cognitive abilities is not new. Studies about the origins of intelligence and attempts to artificially produce intelligence were carried out long before the birth of artificial intelligence and robotics.

Indeed, understanding the mind and the origins of intelligence has been one of the first objectives of philosophers and psychologists. The different schools of thought and ideas of important thinkers inspired engineers and scientists in the attempt to produce machine intelligence. In this respect, one of the most influential thinkers was surely Descartes, prominent responsible for the theoretical duality of mind and body. Sensation requires the physicality of the body; human reason and judgment require the autonomy of the soul. Human beings may have no choice about how the world appears to them, but they can step back from appearances, and allow reflection and judgment.

The Cartesian view strongly influenced other disciplines and approaches to the study of intelligence, among which the *cognitivism* approach and, subsequently, the classical artificial intelligence approach. *Cognitivism* is a theoretical framework for understanding the mind. Its first tenet is that the central function of mind consists of discrete, internal mental states (representations or symbols) manipulated by rule-based transformations. Classical artificial intelligence can be defined as the attempt to produce machine intelligence by methods that reflect Cartesian and *cognitivist* attitudes, first and foremost the centrality of the concept of representation. Following these principles, the first attempts to develop intelligent machines were strongly based on planning and symbolic manipulation.

The failure of the classical approach in developing machines that could cope with the real world brought a gradual shift in focus, from Descartes' mind-body duality to a view of intelligence as embodied and situated. The new view suggests that thinking beings should be considered first as beings acting in their environments. This shift gave rise to a new approach to robotics, characterized by fully reactive robots and the absence of symbolic reasoning, the so called reactive robotics (Brooks, 1999). Since then, reactive robotics and planning have always been considered as two antithetical ways of developing robotics systems. In the remainder of this chapter, we will present in more detail the pioneering works in the history of planning and robotics and finally we will describe some of the works that have attempted to endow robot swarms with simple, cognitive abilities.

2.5.1 Cognition and planning in robotics

The 1950s were a key decade for both artificial intelligence and robotics. In 1956, two important meetings were organized that marked the affirmation of artificial intelligence. During the Symposium of Information Theory, Newell and Simon (Newell and Simon, 1956) presented a program that could demonstrate theorems in logic, while the Dartmouth Conference led to the definition of the "brain-computer metaphor": the processes of the mind were considered completely logical and thus they could be simulated by a well defined program. It became natural to think of human beings as systems that receive input from the environment (sense), process the information by creating a model (model) and reasoning on it (plan), and act depending on the decision formulated (act). This paradigm, inspired by the Cartesian and cognitivist views, gave importance to mental states only, claiming that every behavior is the result of a planning activity. Robotics, which until then had been limited to the cybernetics approach of reproducing animal behaviors using control theory and statistical information theory (Walter, 1951), was strongly influenced by the new sense-model-plan-act paradigm. Robots became machines that reason on a symbolic representation of their world to produce a plan to be executed. This approach is at the basis of two prominent projects developed at the beginning of the 70s, which influenced the research in robotics and artificial intelligence for the following 20 years.

The first project was carried out at MIT and called *The MIT Robot* (Winston, 1972). In this work, a vision system and a robotic arm were programmed to perceive the arrangement of a set of blocks and to build a copy of the structure with additional blocks. The main focus of the project was on computer vision and the goal was to show that a complete three-dimensional description of the world could be extracted from an image. The role of AI was to take this description of the world and manipulate it to formulate plans using rules that describe how the world works.

The second project was carried out at the Stanford Research Institute and called *Shakey the robot* (Nilsson, 1984). Shakey was placed in an environment consisting of a set of rooms. The primary sensor of Shakey was an on-board black-and-white television camera. An off-board computer analyzed the images captured by the camera and merged descriptions of what was seen into an existing, symbolic model of the world. A planning algorithm, STRIPS (Fikes and Nilsson, 1972), operated on those symbolic descriptions of the world to generate a sequence of actions for Shakey. Depending on the goal given, Shakey would navigate around obstacles consisting of large painted blocks, push them out of the way, or push them to some desired location.

Even though both projects showed promising results, they were successful only thanks to a carefully engineered setup: the solutions adopted by the MIT robot were very specific to the world of blocks with rectangular sides and would not have worked in the presence of simple curved objects or under different light conditions. Similar issues affected Shakey, which operated in empty rooms except for the large colored blocks, with walls of a uniform color and carefully lighted. Moreover, in both experiments the environment was mostly static and the result of the actions completely predictable. These issues, and in general the inability of these systems to cope with real-world (unpredictable) scenarios, brought soon the attention of the AI community to the fundamental problems of the classical approach: the frame problem and the symbol grounding problem.

The frame problem (Dennett, 2006) arises from the inability to consider all possible side effects of the execution of a certain action. This stems from the assumption in logic that in the modeled system everything stays unchanged but the direct consequences of the action performed. A possible solution would be to consider all implications of any action before performing it. However, this would not be a viable solution in the real world, as the environment would change while the robot is busy considering all the (infinite) cascade of implications of a certain action. Thus, the frame problem highlights the difficulty of symbolic systems to take into account all changes in the environment and consequently update the model: the more the environment is dynamic and unpredictable, the more the model will fail in tracing changes in it.

The second problem of classical systems is even deeper and harder to address: the symbol grounding problem. In order to be effective in the real-world, a symbolic system that reasons must have an understanding of what it is reasoning about. However, symbolic systems reason using only syntactic rules, while ignoring the semantics. Therefore, understanding does not pertain to such systems: the rules of reasoning used by symbolic systems were formulated by human beings, observers who already have those symbols grounded—i.e., given a real-world meaning. For a complete discussion on symbol grounding and other issues related to symbolic systems we refer to Anderson (2003).

Due to numerous problems of the classical approach, different researchers began, around the mid-80s, to rethink the origin of intelligence. They all identified the main issue of classical systems in the lack of ability to promptly react to contingencies due to the over-commitment to reasoning and planning. Kaelbling and Rosenschein proposed the concept of situated automata (Kaelbling, 1987; Kaelbling and Rosenschein, 1990). The architecture of an agent is divided in two components: the perception component produces the model of the world in input to the action component, which in turn maps this information to action. The action component is specified as a set of pairs of the type condition-action. The idea is not to wait until a plan has been completely formulated, but rather get the information produced at every cycle and map it directly into action. Schoppers introduced the concept of universal plans (Schoppers, 1987). The novelty respect to the classical approach stands in the fact that the representation used specifies appropriate reactions for every possible situation in a given domain. In this way there is no commitment to a particular course of actions. At design time, the offline planner partitions the set of possible situations on the basis of the reaction that every situation requires. At operation time, the actual situation is classified and the response planned for that class is performed. Sanborn and Hendler (Sanborn and Hendler, 1988) proposed a system equipped with a reactive subsystem that was designed to keep the reasoner out of trouble. The reactive subsystem keeps a reduced database of facts that are relevant only for the near future of the agent. A component called monitor repeatedly evaluates some aspect of the immediate actual situation and determine whether there is a discrepancy within the current short term database. If a discrepancy is detected, the monitor fires inhibiting or enabling different actions of the agent. Instead, as long as the short term database is stabilized and holds, the actions are selected by the long-term reasoner. Georgeff and Lansky attempted to merge a classical planner with a reactive system able to guarantee the survival of the agents in highly dynamic worlds with a system called procedural reasoning system (Georgeff and Lansky, 1987). The procedural reasoning system is similar to a traditional planner but it provides reactivity to dynamic changes of the environment. The system chooses the best procedure to execute in the particular moment, but during the execution new information about the world can be produced that changes the goal of the agent and causes the reassessment by the system of the best procedure to execute.

These works showed a first transition from systems completely committed to planning and symbolic reasoning towards systems with reactive components. The real revolution, however, began with Brooks (Brooks, 1991, 1986, 1999). Brooks introduced a new approach characterized by fully reactive robots and the complete absence of symbolic reasoning. Given the problems expressed by the classical approach, different researchers seemed to agree on the solutions: shorter plans, more frequent attention to the environment, and selective representation. Brooks pushed these solutions to the extreme: the logical end of shorter plans is not to have a plan; likewise, the limit of more frequent attention to the environment is constant attention. Finally, selective representation leads to closing the gap between perception and action. Brooks' ideas led to the so called *reactive robotics*. Research in reactive robotics does not seek to produce a human-like thinking process. Instead of focusing on machines that can think intelligently, the emphasis is on the creation of machines that can act intelligently. In order to act, the robots continuously refer to their actual perception of the environment rather than to an internal model of the world. Two key concepts in reactive robotics are *situatedness* and *embodiment*. Situatedness refers to the fact that robots are immersed in their world, which can be directly perceived through their sensors (sense), without the need of any modeling stage. This solves modeling problems such as the frame problem. Embodiment refers to the fact that robots act in the world (act). Through their actions, robots modify the world and the feedback that they will receive from it. Brooks claimed that this tight coupling between the robots and their world provides a starting point to ground concepts, that is, to give concepts a real-world meaning. In his view, a symbol can acquire meaning only if grounded by the experience of the world. Therefore, embodiment solves the symbol grounding problem through the interaction of the robot with its environment.

The new reactive approach produced immediately striking results and the work adopting the new paradigm flourished. However, the first critics came soon. Maes (1990) recognized a shared idea in these new architectures, the concept of "emerging functionality" from the interaction of the robots with their environment. An important implication of this property is that it is not possible to directly tell the robots how to achieve a goal. Instead, it is necessary to find a dynamics, or interaction loop, that involves the system and its environment which will eventually converge toward the desired goal. This solution requires the designer to hard-code or pre-compile the action selection, as the robots lack explicit goals and goal handling capabilities. The author proposed to merge the advantages of classical and reactive approaches in order to obtain a robust, flexible and fast action selection. The interesting aspect is that the approach used to achieve planning abilities is completely different from the classical one: action selection is achieved in an implicit fashion. It emerges in a distributed way by parallel interactions between a set of simple modules that exchange a virtual form of energy through different types of links: successor links, predecessors and conflict links. The energy spreads through the network starting from the current situation (actions are given energy if they match the current situation) or from the goal if an action achieves a goal. Similar critics to the complete absence of representation and a similar approach to Maes' one were raised by Mataric and Brooks (1990) and Mataric (1992) to obtain a representation of a map from the interactions of reactive behaviors of a single robot. The authors also showed how, with the same mechanism, it was possible to obtain a path to follow on that map representation.

Excluding the works by Maes and Mataric, who tried to deeply combine the two approaches, the rest of the robotics community has since then considered the classical and the reactive paradigm as two antithetical ways of developing robotics systems. Even though examples of hybrid models exist (Arkin, 1990; Saffiotti et al., 1995), the two

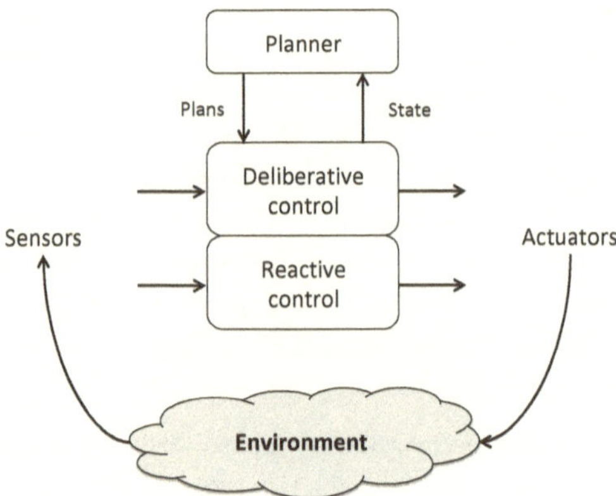

Figure 2.3: **Hybrid system architecture for single robot.** A planner module receives the state of the environment as perceived by the sensors. It formulates long-terms plans and passes them to a deliberative control subsystem. The deliberative control subsystem, together with the reactive control subsystem, selects the behavior of the robot.

aspects remain clearly distinct from the logical point of view. A robot that adopts an hybrid model, in fact, alternates phases in which it plans the next sequence of actions, with phases in which it performs these actions while possibly reacting to contingencies. The two phases are clearly separated at the level of the robot control software architecture (see Figure 2.3): A reactive component drives the robot and deals with short-term decisions, while on a higher level a planner reasons about long-term goals and selects, through different mechanisms and policies, the appropriate reactive behavior to execute. The same distinction and the same principles were applied to multi-robot systems, starting from the late 90s. Some system uses a central planner to compute the different subplans that each robot should perform (Jensen and Veloso, 1998) while in fully distributed systems each robot is equipped with its own planner (Matellán and Borrajo, 1998; Guzmán-Alvarez et al., 2013; Goldberg et al., 2003; Magnenat et al., 2009; Spaan et al., 2006). In both cases, an additional phase of coordination and synchronization among the robots is required in order to avoid failures and keep plans and execution consistent.

Figure 2.4 shows a system with a central planner, Figure 2.5 a system with distributed planning abilities.

After Brooks, nobody has tried to propose a new paradigm for developing robotics systems. All the works carried out since the late 90s have focused on slightly improv-

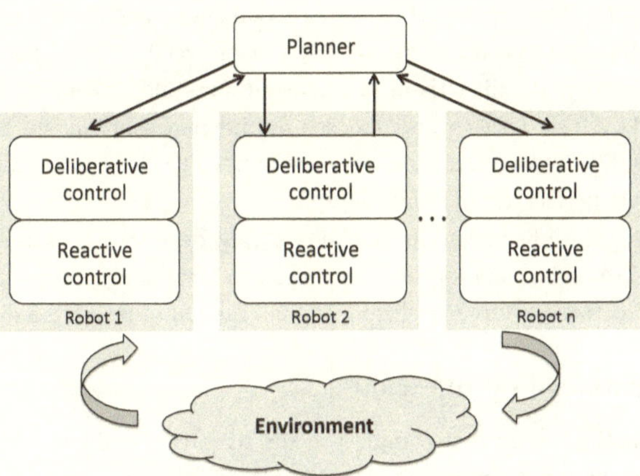

Figure 2.4: **Hybrid system architecture for multi-robot systems with a central planner.** The central planner formulates the long-term plans for all the robots.

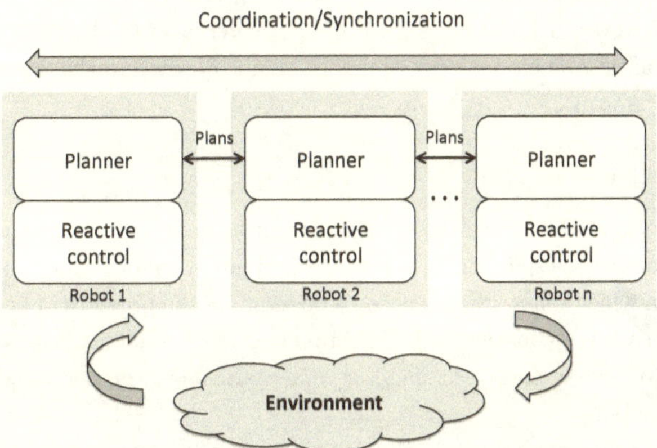

Figure 2.5: **Hybrid system architecture for multi-robot systems with distributed planner.** Every robot has its own planning subsystem. The robots need to coordinate and synchronize their plans in order to work towards the same goal.

ing one or few aspects of the same architecture. Some focused the attention on the planning technique, proposing fuzzy techniques (Saffiotti et al., 1995) rather than hierarchical planning (Magnenat et al., 2009) or techniques based on decentralized partially observable Markov decision processes (Spaan et al., 2006). Some others gave different roles to the robots, dividing them in planner robots and executor robots (Guzmán-Alvarez et al., 2013). Other works focused on different solutions for the coordination of the robots, like for instance marked-based approaches for the allocation of subtasks to different robots (Gerkey and Mataric, 2002).

The idea that we formulate in this **book** is that deliberative (sense-model-plan-act) and reactive (sense-act) approaches can coexist in a particular way: the ability to plan can emerges at the collective level from the interaction of reactive individuals.

2.5.2 Cognition in swarm robotics

In swarm robotics, each robot complies to the reactive paradigm (sense-act): at each control cycle, the robot senses the environment and acts accordingly, without relying on representations of the environment or symbolic reasoning. Complex collective behaviors of the swarm emerge from large number of individual-individual and individual-environment interactions.

Designing individuals so that their cooperation will engender the desired collective behavior is a difficult endeavor (see Section 2.2). For this reason, most research has so far produced robot swarms that display spatial/geometrical properties or mechanical abilities. However, a few studies have been devoted also to the emergence of simple cognitive abilities. The remainder of this section is devoted to the description of these latter robot swarm.

Fault detection

In collective fault detection, the swarm recognizes faulty robots and initiates appropriate responses. Despite being robust to individual robot failures, there might be situations in which robot swarms need to be aware of the presence of faulty robots and react properly. We consider fault detection a cognitive ability because it involves a form of introspection and understanding of a swarm's own property (i.e., the presence of faulty robots).

In a pioneering work on collective fault detection, the robots use an algorithm inspired by firefly synchronization (Christensen et al., 2009). The robots emit a periodic signal and eventually synchronize with each other using a model of pulse-coupled oscillators (Izhikevich, 1999). If a robot does not synchronize with its neighbors, it is assumed to be faulty and a response can be triggered. A more recent approach takes inspiration from the functioning of cells of the immune system to develop a robust fault-detection algorithm (Tarapore et al., 2017). The authors then demonstrated the effectiveness of their approach in a set of physical-robot experiments (Tarapore et al., 2019).

Fault detection and fault tolerance are particularly critical when the target mission of the swarm requires extensive communication between the robots. In these cases, even the presence of a few *Byzantine* robots—that is, faulty or malicious robots—may be sufficient to make the entire mission fail. A novel approach proposes the use of blockchain technology to address the security issue produced by the presence of Byzantine robots (Strobel et al., 2018). The approach was tested on a decision-making task in which the robots have to reach a consensus on the most present tile color in an environment in which the floor is covered with black and white tiles. The results show that the blockchain-based approach is able to effectively detect and exclude Byzantine robots and thus greatly improves the overall fault tolerance of the swarm.

Group size regulation

Group size regulation requires a similar form of introspection to that of fault detection. In group size regulation, the robots have the ability to estimate and regulate their number in a group. This ability is useful, for example, when an excessive number of robots in a group lowers the performance of the swarm. One of the first solutions proposed to estimate the number of robots in a group takes inspiration from the behavior of fireflies (Melhuish et al., 1999a): the robots emit a signal at random times and count the number of signals perceived over a period. This number is used to estimate and tune the size of the group. An improvement of this algorithm based on a more strict signaling order can obtain more reliable group size estimates (Brambilla et al., 2009). In other studies a set of flying robots aids the aggregation of ground robots (Pinciroli et al., 2010). The aerial robots estimate the size of the aggregate and communicate to the ground robots the accordingly adjusted probabilities of joining or leaving the aggregate. Through this mechanism, the ground robots are able to form groups of different sizes. Finally, a study proposed an algorithm inspired by cockroaches aggregation under shelters that is able to partition the swarm in groups of different sizes (Pinciroli et al., 2013). The aggregation in different groups develops in parallel, therefore the convergence time of the algorithm is independent of the number of groups.

Collective perception

In collective perception, the robot swarm leverages its redundancy to acquire information on one or more characteristics of the environment.

One of the first work towards collective perception used evolutionary algorithms to synthesize a robot swarm able to determine the density of black spots on the ground (Morlino et al., 2012).

A more recent work compared different strategies (majority rule, voter model, and direct comparison) to develop a robot swarm able to determine the prevalent color of the ground (between two alternatives) (Valentini et al., 2016a). Experiments were run with a swarm of up to 20 e-puck robots.

Collective decision-making

Collective decision making focuses on how a robot swarm can make decisions. Different categories of situations can require a swarm to make a choice: The first category comprises situations in which the robots need to reach a consensus on a single choice among a set of possible alternatives. The behaviors that aim at solving these problems are called *consensus achievement*. Often, consensus achievement requires forms of collective perception to then make a decision based on the perceived environmental features. The second category comprises problems in which the robots have to distribute themselves among a set of possible tasks and operate in parallel on those tasks in order to maximize the performance of the system. This process is called *task allocation*. The last category we consider in this **book** comprises problems in which the robots need to decide in which order to perform a set of given tasks. This process is called *task sequencing*.

A comprehensive survey on the most effective models and methodologies used to obtain collective decision making can be found in (Hamann, 2018).

Consensus achievement   Consensus achievement behaviors allow a robot swarm to converge on a single choice among a set of alternatives. The choice is usually the one that maximizes the performance of the system. Consensus achievement can be observed in many insect species; for example, ants can determine the shortest of different paths using pheromone (Camazine et al., 2001) and bees collectively choose the best nest location among several alternatives (Couzin et al., 2005).

The robots of a swarm can reach consensus using either direct or indirect communication. Several strategies have been proposed in the literature. By mimicking a form of consensus achievement used by cockroaches, it is possible to develop robot swarms that choose a single aggregation zone using indirect communication (Garnier et al., 2005)—that is, decisions are taken using indirect clues, such as the density of neighbors. Consensus achievement can be achieved via quorum sensing, an algorithm inspired by the choice of the best over N alternatives in ants and bees (Parker and

Hong, 2011).

The algorithm is based on direct communication: robots evaluate alternatives and advertise them through recruiting messages that are broadcast with a frequency proportional to their perceived quality. This allows the swarm to eventually converge on the best alternative. The nest-site selection in honeybee colonies inspired other work: a thorough study of its analytical model and the identification of the parameters that determine its working regime enabled the definition of guidelines for the implementation of the individual robot behavior. This approach was applied to the shortest path selection problem (Reina et al., 2015a).

Consensus achievement has been used also to let the robots agree on a common reference orientation (Navarro and Matía, 2012). The algorithm, which uses only relative positioning and local communication, can be used as a preliminary step for collective motion.

Two novel strategies were proposed: the first is based on a weighted voter model and ensures a high decision accuracy and robustness to noisy assessments of alternatives (Valentini et al., 2014); the second couples a mechanism of time-modulation with individual robots' decisions based on the majority rule and has been shown to speed up the decision process considerably (Montes de Oca et al., 2011; Valentini et al., 2015b,a). A later work (Scheidler et al., 2016) introduced the k-unanimity rule, which extends the control structure of the majority rule with differential latency: A robot using the k-unanimity rule changes its current opinion to a different option only after consecutively encountering k other robots all favoring that other option. This model was studied with up to 3 options. Another work focused on the adaptability of decision-making strategies to abrupt changes in the qualities of the options (Prasetyo et al., 2019). The work compares two adaptation mechanisms called stubborn agents and spontaneous opinion switching. With the aid of modeling techniques and computer simulations, the authors show that both strategies can achieve adaptability to dynamic environments, and identify the parameters that regulate the level and the speed of adaptation of the robot swarm. Consensus achievement has been used to allow a robot swarm to decide when it is time to switch the the next task of a known sequence of tasks (Parker and Zhang, 2010). The problem of deciding whether to switch to the next task is a binary decision-making problem whose options are "task complete" or "task incomplete".

A pioneering work (Strobel et al., 2018) proposed the use of blockchain technology to increase robustness of consensus-achievement strategies to the presence of Byzantine robots—that is, faulty or deliberately malicious robots that may cause the entire swarm to fail its mission.

Consensus achievement is typically modeled using Markov chains (Valentini et al.,

2013; Valentini and Hamann, 2015; Hamann, 2013; Valentini and Hamann, 2015). Another modeling technique used to model consensus achievement is based on Bio-PEPA (Massink et al., 2013), a process algebra originally introduced for modeling biochemical systems. From a formal specification of the system in Bio-PEPA, the authors were able to analyze the behavior of a swarm that had to identify the shortest path between two possible choices.

For a comprehensive review of the literature on consensus achievement we refer to Valentini et al. (2017).

Task allocation Task allocation behaviors deal with the distribution of robots over a set of different tasks. The allocation is usually dynamic and aims at maximizing the overall performance of the swarm. Ants and bees use task allocation. For example, while some individuals are foraging, a number of other individuals are in charge of looking after the larvae. Task allocation is usually obtained through manual design methods. In the first works on task allocation the choice of the robots was limited to whether to engage in a foraging task or remain in the nest, resting. The robots would choose on the basis of the level of energy in the nest (level raised by the preys collected and consumed by robots resting) (Krieger and Billeter, 2000), or on the basis of individual observation of the environment and of other robots (Agassounon and Martinoli, 2002). A similar approach is used in other works to design task allocation for two sequentially interdependent tasks (Pini et al., 2013; Brutschy et al., 2014). The process is based on individual observations of the environment and of the current performance of the swarm, and hence it does not require direct communication between robots. Other works focus on whether, when, and how the robots should perform the overall task or partition it and allocate themselves to one of the subtasks (Pini et al., 2011). For example, a robot could choose whether to carry a prey from the source to the nest or store it in a cache (Brutschy et al., 2015). In the latter case, the prey would then be collected by robots waiting on the other side of the cache and carried to the nest. The choice is made on the basis of the estimation of the costs involved.

Some attempts have been made to design task allocation via automatic methods. A study demonstrated how self-organized task allocation can be obtained through evolutionary robotics (Ferrante et al., 2015). Differently from the previous attempts that used artificial neural networks (Tuci, 2014; Nitschke et al., 2012b,a), the authors used a method of grammatical evolution to automatically design task allocation strategies. This method was proven more effective, and was able to successfully design task allocation both when provided with few behavioral building blocks and when the strategy had to be defined anew.

Task allocation is usually tested on foraging. However, there are examples of application of task allocation to other practical problems. The allocation of robots to different operations on a construction site (Yun et al., 2011) and a stick-pulling problem (Halász et al., 2013) are two further examples of application.

Often, task allocation is used as a testbed for modeling techniques. A set of advection-diffusion-reaction partial differential equations was used to model and design a behavior for task allocation (Berman et al., 2009). Delay differential equations were used to model task allocation and verify the stability of the system obtained (Hsieh et al., 2008). A population dynamics model was used to determine some parameters of the individual behavior and the optimal distribution of robots in two task-allocation scenarios (Correll, 2008). Finally, a Poisson process was used to describe the performance of a swarm performing a set of tasks and to design the individual switching probabilities for task allocation (Khaluf et al., 2014).

Task sequencing

Although the composition and sequencing of tasks to create more complex behaviors has been widely adopted in the single-robot case (Arkin, 1998; Mataric, 1998; Belta et al., 2007), only few works have attempted to use the same approach for robot swarms.

Parker and Zhang (2010) proposed a consensus-achievement algorithm to let the swarm take an autonomous decision on when to swicth to the next task. This work is limited by the fact that the robots decide *when* to switch and not *what* to do next—i.e. the correct sequence of the tasks is known at design time and is hard coded in the robots' control software.

A step towards task sequencing in robot swarms was performed in Nagavalli et al. (2017): An off-line algorithm is used for sequencing the set of collective behaviors (selected from a larger set of known and well-defined possible behaviors) that will lead the robot swarm from an initial state to a goal state. The switch between behaviors can occur only at particular decision points and between two decision points all the robots of the swarm execute the same set of rules. This means also that the individual rules that the single robots execute are pre-defined so that the selected collective behavior can emerge. The algorithm is proven complete and optimal, but the execution times are too high for real-time control of a real robot swarm. The algorithm was applied to a navigation scenario, in which the swarm has to navigate between a starting and a goal position while avoiding obstacles, and an area-coverage scenario. Experiments were executed in static environments and simulation only.

In the remainder of this **book**, we present a robot swarm with the ability to sequence task autonomously at operation time.

Chapter 3

TS-Swarm Mark I

In this chapter, we present TS-Swarm (Fig. 3.1). Differently from most of the robot swarms that have been demonstrated in the literature, TS-Swarm displays a complex cognitive ability: the ability of sequencing tasks. It does so despite the fact that the individual robots of the swarm act by reacting to contingencies and are not aware of the task-sequencing problem being solved. Although the reactive paradigm was shown to be more effective for implementing the behavior of individual robots (Brooks, 1991), we argue that robot swarms require complex cognitive abilities to solve the unpredictable problems posed by real-world applications. The robots of TS-Swarm thus optimize online their collective behavior to solve the task-sequencing problem that allows them to accomplish the given mission.

To sequence tasks while individually behaving reactively, the robots of TS-Swarm form a chain structure that ultimately serves two purposes: (i) assisting the navigation of other robots between tasks; and (ii) encoding the order in which tasks must be performed.

The order in which the tasks must be performed is discovered collectively by the robot swarm via a trial-and-error process: when a robot acquires additional information about the order of execution of a task, it shares this information locally with the chain member near the location where the task is performed. From then on, the chain member will use this information to assit other nearby robots in performing the task in the correct order.

In this chapter, we present the first version of the system: TS-Swarm Mark I (Mark I)[1]. In Mark I, a robot acquires more information about the correct order of execution of a task by executing it and immediately receiving a negative feedback if it was performed in the wrong order, positive feedback otherwise. A robot receives

---

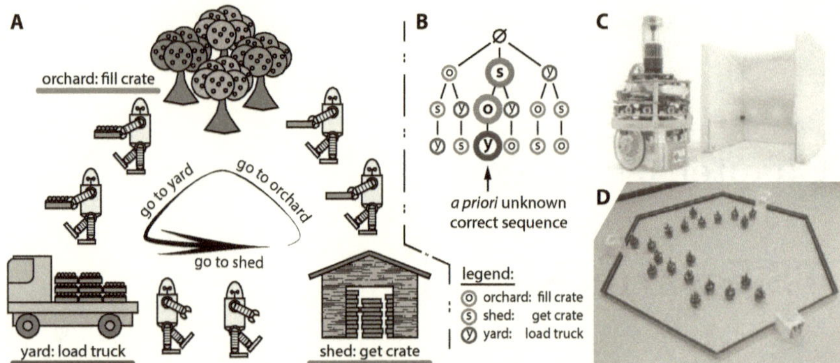

Figure 3.1: **From task-sequencing missions to TS-Swarm.** (**A**) The example of task-sequencing mission presented in Fig. 1.1, a fruit-picking robot swarm. The correct sequence of the three tasks must be repeated multiple times to fully load the truck with fruit. (**B**) Formal representation of the solution space. (**C**) An e-puck and a TAM (see Fig. 3.2 and Appendix A). (**D**) TS-Swarm in its arena with three TAMs.

feedback in the sense that, after performing a task, it becomes immediately aware of whether the task was performed in the correct order or not.

The rest of the chapter is organized as follows: First, in Section 3.1, we describe the abstract task-sequencing algorithm at the base of Mark I. Then, in Section 3.2, we introduce the hardware and software platforms on which we implement this algorithm. In Section 3.3, we describe the details of the implementation of Mark I, focusing mostly on the case in which the tasks to sequence are three (Mark I$_3$), and then presenting the minor modifications that we made to obtain a system able to sequence four tasks (Mark I$_4$). In Section 3.4, we present the experiments that we performed to demonstrate the task-sequencing ability of Mark I and we analyze the results obtained. Finally, in Section 3.5, we discuss some possible improvements to the system.

3.1   Distributed task-sequencing algorithm

In this section, we describe the task-sequencing algorithm at the base of Mark I. To do so, we consider the case in which the tasks to be sequenced are three, but all the considerations we make here are general and applicable to the case in which the tasks are m. The three (m) tasks must be performed in a specific order (without repetitions) by an individual robot of the swarm. Each task must be performed in a certain area, and the correct order is unknown at design time. The sequence of tasks must be repeated multiple times by the same or by other robots.

In TS-Swarm, all robots follow the same reactive rules but autonomously assume different roles depending on the contingencies they encounter. A robot can be a *runner*, *guardian*, *tail*, or *link*. We will collectively refer to guardians, links, and tail as *chain members*. Initially, all robots are runners and move randomly in the arena. Upon encountering a task, a runner performs it and then positions itself in the proximity of the task, becoming its guardian. From then on, no other runner will perform the task, unless directed to do so by its guardian. Eventually, three robots are the guardians of the three tasks. Two of them received negative feedback, as their task is not the first of the sequence. The other guardian received positive feedback: its task is the first one.

Consider the fruit-picking example described in Chapter 1 and illustrated in Fig. 1.1 and Fig. 3.1. The robots composing the fruit-picking swarm are pre-programmed to perform the individual tasks—i.e., pick up a crate, fill the crate with fruit, load the crate onto the truck—but initially ignore the correct order of execution of the tasks and the location of the area where each task must be performed. The first robot reaching the area where a task must be performed attempts to perform it. In this example, the first robot that reaches the shed reacts to the presence of crates by trying to pick up a crate. This will result in a success (positive feedback), as the robot did not have anything in its hands. On the other hand, the first robot that reaches the orchard reacts to the presence of fruit on the trees by trying to pick the fruit and place it in a crate. This will result in a failure (negative feedback), as the robot does not yet have a crate to fill with fruit. Lastly, the first robot that reaches the yard reacts to the presence of the truck but fails to load it (negative feedback), as it reached the yard empty-handed. After attempting to perform the task, each of these three robots positions itself in the proximity of the task and becomes its guardian.

The guardian that succeeded in performing its task—i.e., received positive feedback— initiates the construction of the chain. Runners that encounter the chain being built can contribute to its extension by positioning themselves at its end, one after the other. We refer to the last robot in the chain as the tail and to the others as the links. Tail and links align and keep a target distance between themselves so that the chain is stretched and straight. If the chain reaches a wall, it turns, sweeping the environment. By extending and turning repeatedly, the chain eventually encounters another guardian. When this happens, the tail transfers its role to the guardian and becomes a link. The guardian initiates the construction of a new branch of the chain to ultimately include all the guardians. Robots that have not become chain members remain runners and navigate the environment following the chain. When a runner reaches a guardian, it performs the guarded task if so directed.

Guardians learn to direct runners via trial and error. As mentioned, a guardian received positive or negative feedback, depending on whether its task is the first to be performed or not, respectively. The guardian that received positive feedback will direct to its task the runners that have not yet performed any task. The other two guardians learn the correct policy with the help of the runners. The first time they are reached by a runner that has performed exactly one task (in our example, a runner that has already picked up a crate), they ask it to perform their task and wait to be informed of the feedback. If the feedback notified by the runner is positive, from then on the guardian will direct to its task the runners that have performed one task. In our example, this is what happens at the orchard, where a runner with an empty crate successfully fills it with fruit and informs the guardian of the positive feedback received. If the feedback notified by the runner is negative, the guardian will direct to its task the runners that have performed two tasks. This is what happens at the yard, where a runner fails to load the truck, as its crate is still empty, and notifies the guardian of the negative feedback received. From then on, the guardian at the yard will direct to its task only runners that have already performed two tasks—i.e, pick up a crate and fill it with fruit.

Finally, the swarm has autonomously found the correct sequence. All the runners will now be able to reliably navigate between the areas using the chain and will be directed by the guardians to perform the three tasks in the correct order. The sequence of tasks will be repeated multiple times by the runners of the swarm to ultimately complete the mission.

3.2   Platforms

We implemented TS-Swarm on e-puck robots (Mondada et al., 2009) and we use a custom-designed device called TAM (task abstraction module) to abstract tasks (Brutschy et al., 2015). As a development environment and robot simulator we used ARGoS (Pinciroli et al., 2012). Here we provide an overview of these platforms. The work carried out for this **book** produced a major contribution to their development. In particular, we re-designed and implemented the e-puck software architecture, and implemented the robot-TAM communication system on the TAM. The implementation of the simulated version of both platforms in ARGoS was also carried out during this work. Overall, the work on the platforms required about one year. The detailed descrition of the contributions and the final platforms is given in Appendix A.

3.2.1 The e-puck

The e-puck is a mobile two-wheeled differential-drive robot designed for education and research (Mondada et al., 2009). It is cylindrical in shape, with a diameter of 70 mm and a height of 50 mm. Its basic version is equipped with a PIC microcontroller and several sensors and actuators. The sensors are: 8 infra-red transceivers, which can be used to sense the presence of obstacles or measure the intensity of ambient light; a col [B] or camera at the front of the robot; a microphone; and a 3-axis accelerometer. The actuators are: two stepper motors, which control the motion of the robot by differential steering (one motor for the left wheel and one for the right wheel); a ring of 8 red LEDs; and a speaker.

The e-puck can be enhanced by the addition of various extension boards. For the research presented in this **book**, we extended the basic version of the e-puck with a range-and-bearing board (Gutiérrez et al., 2009), an omnivision module, and a Gumstix Overo board (Figure 3.1C, left). The range-and-bearing board enables local communication between e-pucks via infra-red signals. It comprises 12 emitters and 12 re-ceivers placed all around the body of the e-puck. The range-and-bearing board allows e-pucks to send and receive four-byte messages at a rate of about 30 messages per sec-ond. Upon reception of a message, the board computes the distance (range) and angle (bearing) of the peer e-puck that sent the message. The omnivision module comprises an omni-directional camera and 3 RGB LEDs and enhances the perception and local communication capabilities of the e-puck. Through the camera, an e-puck can see its neighboring peers and the TAMs. Moreover, it can perceive the color coded status that the neighboring peers might display using their RGB LEDs. The Gumstix Overo board increases the computational capabilities of the e-puck and provides the flexibility and potential of a computer running Linux. It allows running C++ code, which is not possible on the PIC micro-controller of the basic version of the e-puck.

The basic version of the e-puck is powered by a rechargeable lithium-ion battery with 5 W h capacity. The omnivision module houses a second battery with the same capacity to cope with the higher energy requirements of the extended e-puck. In a typical experiment, the full battery charge of an extended e-puck lasts about 40 minutes. Indeed, we have observed that after about 45 minutes of continuous operation, the charge of the batteries is low. This negatively affects the behavior of the robots and in particular their ability to successfully transmit and receive messages through the range-and-bearing board.

3.2.2　The TAM

The TAM (Brutschy et al., 2015), task abstraction module, is a device conceived for facilitating laboratory experiments with e-puck robots. A TAM represents an abstract task to be performed by an e-puck. The goal of the TAM is to abstract from task-specific details that are irrelevant to the objectives of an experiment. The TAM is particularly useful in experiments that focus on group dynamics rather than on the specific tasks performed by the individuals.

The TAM is a booth that an e-puck can enter. For an e-puck, being into a TAM for a given time span amounts to performing the task abstracted by the TAM itself. The TAM has a cubical shape with sides of 120 mm (Figure 3.1C, right). The TAM is controlled by a microcontroller (ATmega-1284p, 16 MHz), and is equipped with two light barriers, three RGB LEDs, and an IR transceiver for short-range communication. Each TAM is powered by a rechargeable lithium-ion battery of 5 W h capacity, the same battery used by the e-puck. In a typical experiment, a full battery charge lasts for over 10 hours. Each TAM is equipped also with an XBee mesh networking module, which enables communication and coordination with a central computer and with other TAMs. A group of TAMs can be therefore programmed to represent complex relationships between tasks. For example, a task could become activated only upon completion of another one or a group of tasks could be performed successfully only in a specific order. The experimenter implements the logic that defines the relationship between tasks on a central computer. The computer dispatches commands to the TAMs to realize the relationships programmed by the experimenter. The TAMs and the central computer communicate wirelessly via the XBee mesh networking module.

An e-puck perceives the colored LEDs of the TAM using its omni-directional camera. Different tasks are signaled by using different LED colors. An e-puck can decide to perform the task represented by a TAM by moving into it. The TAM detects the presence of the e-puck by means of its light barriers and reacts according to a logic defined by the experimenter. For example, upon the detection of an e-puck, the TAM could change the color of its LEDs or start communicating with the e-puck itself. The TAM and the e-puck communicate with each other through their infra-red transceivers. Communication between e-pucks and TAMs enables experiments in which e-pucks receive individual feedback for the tasks they perform.

In the framework of this book, we contributed to the development of the TAM by designing and implementing the robot-TAM communication subsytem, and by implementing the coordination software running on the central computer.

3.2.3 ARGoS

ARGoS (Pinciroli et al., 2012) is a modular multi-robot simulator and development environment conceived for being flexible and efficient. ARGoS provides a straightforward way to port control software developed in simulation to the robots, without requiring any modification. To achieve this result, each sensor and actuator presents an interface with two back-end implementations: one for simulation and one for the robot. The control software of the robot directly interacts with this interface without having knowledge of which back-end implementation is being used. At link time, ARGoS takes care that the appropriate back-end implementation is executed, depending on whether the execution is to take place in simulation or on the robot. ARGoS provides a number of physics engines. Some of them are kinematic engines that favor performance over realism, others are dynamics engines, in two or three dimensions, that require more computation but that produce more realistic simulations. As realism plays an important role in our simulations and the system we propose comprises only robots that move on the ground, we use a dynamics engine in two dimensions in all the simulated experiments.

In the framework of this **book**, we contributed to the development of ARGoS by extendeding the basic model of the e-puck with models of the range-and-bearing board, the omnivision module, and the Linux board (Garattoni et al., 2015). The extended e-puck plugin for ARGoS was made available to the research community and it has already been used by several research teams around the world. We also created the model of the TAM, which was not originally provided by ARGoS. We refer the reader to Appendix A for a complete description of the platforms and their development for this book.

3.3 Description of Mark I_m

Mark I_m assumes that a robot receives feedback on the order of execution of a task immediately after executing it. In the following of this section, we describe the implementation of the system. Although all the robots of the swarm execute the same control software, they assume different roles at runtime. Roles are not pre-assigned, but are rather taken autonomously by the robots on the basis of their interactions with peers and environment. We describe each role that the robot can take at runtime—i.e., guardian, link, tail, and runner—providing details of their implementation on e-puck robots.

The control software of the robots is realized as a probabilistic finite state machine in which the control cycle has period of 100 ms. A high-level representation of the

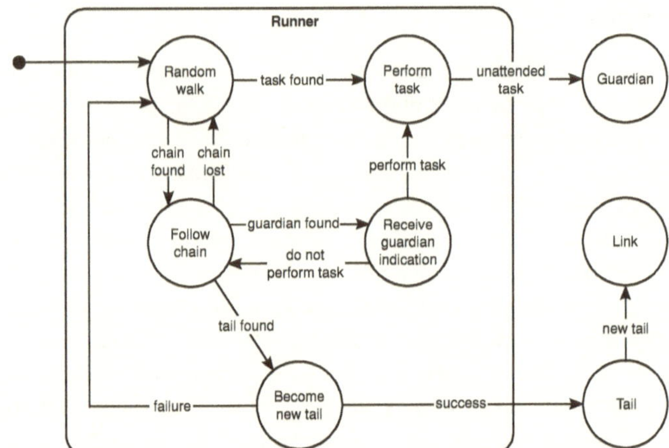

Figure 3.2: **State machine of TS-Swarm.** High-level description of the robots' behavior.

probabilistic finite state machine is given in Figure 3.2.

At every control cycle, all chain members broadcast a four-byte message via their range-and-bearing board. On the other hand, runners broadcast a four-byte range-and-bearing message only in specific situations, as it will be detailed in the following. The general scheme of the message encoding is given in Figure 3.3. Details will be provided on a per-role basis in the following of this section.

In the remainder of this section, we detail the implementation of the four robot roles of Mark I_m for the extended e-puck platform.

Guardian

The guardians are robots that position themselves right in front of a task to signal its presence and to indicate whether a runner should perform it or not. They announce their role by displaying the color cyan through their RGB LEDs (Figure 3.4). The guardians also act as end points of the branches of the chain. In particular, they are in charge of starting the construction of a branch when: (i) the task they guard is the first to be performed, or (ii) they are reached by the chain being built (provided that the whole chain is not completed, yet).

Before becoming a guardian, a robot is a runner that explores the environment performing a random walk. A runner becomes a guardian when it encounters an unattended task, performs it, and eventually stand in front of it. After performing the task of which it becomes a guardian, a robot receives positive feedback if the task is the

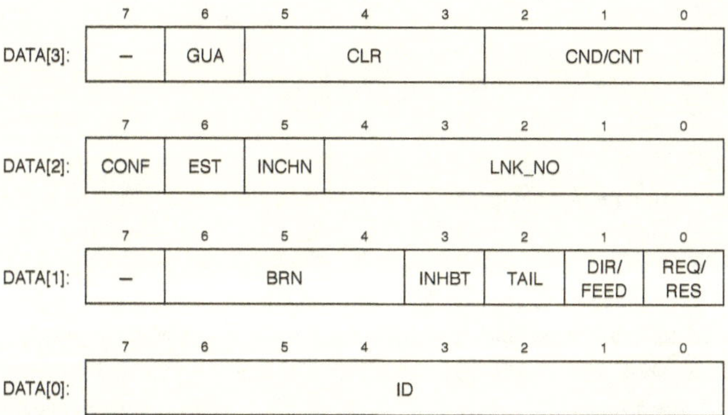

Figure 3.3: **Encoding of the range-and-bearing message in Mark I_m.** Guardians identify their messages by setting DATA[3][6]. Each guardian locally broadcasts its CND and CONF (see Section 3.3-Guardian) by setting DATA[3][2:0] and DATA[2][7], respectively. Additionally, a guardian indicates the color of the guarded task (DATA[3][5:3]), whether it is in chain or isolated (DATA[2][5]), and whether the nearby runners should be inhibited (DATA[1][3])—except the one that initiated the inhibition and whose ID is indicated in DATA[0][7:0]. When it is reached by a branch of chain, a guardian also sets DATA[2][6] to indicate that the branch has been established and increments by 1 the value of BRN (DATA[1][6:4]) received from that branch. BRN enumerates the branches of chain that have been established; it is used to identify each branch and to determine when all the tasks have been connected by the chain. The links along a branch relay the value of BRN indicated by the guardian that initiates the branch. Links indicate also their position within the branch (DATA[2][4:0]) and whether the branch is established or still being built (DATA[2][6]). The tail identifies its messages by setting DATA[1][2]. To indicate the sector in which a new tail can join the chain, the tail sends, only in that sector, a message in which DATA[1][1] is set. To respond to a request for becoming a new tail, the tail sends a response by setting DATA[1][0] and inserting the ID of the selected requesting runner in DATA[0][7:0] (see Section 3.3-Tail and Figure 3.9). Runners' messages are characterized by the value 0 in the fields GUA, TAIL, and LNK_NO. Runners can send a request to join the chain as the new tail by setting DATA[1][0]. They notify a guardian of the success/failure of a task execution by indicating their value of CNT in DATA[3][2:0] and the feedback received in DATA[1][1] (see Section 3.3-Guardian and Figure 3.6). Finally, a runner can inhibit other neighboring runners while communicating with a guardian or performing a task by sending a message in which DATA[1][3] is set (see Section 3.3-Guardian and Section 3.3-Runner).

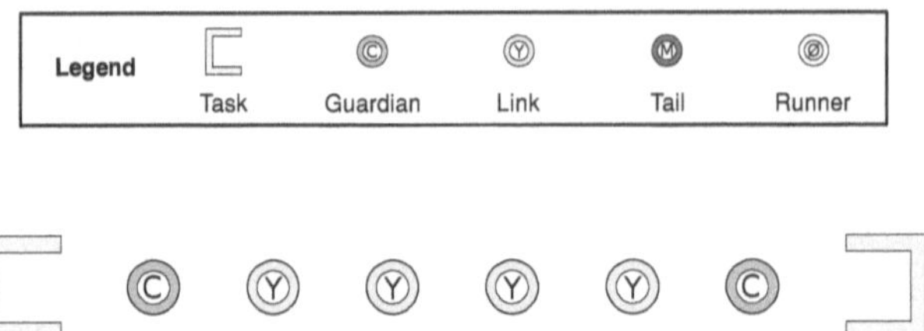

Figure 3.4: **Guardians.** The guardians are the chain members that are at the two ends of each branch. They are identified by the color cyan. They place themselves right in front of a task and instruct other robots on whether they should perform it or not.

first one to be performed and a negative one otherwise. More precisely, to act as a guardian, a robot positions itself so that the TAM of the guarded task is right behind its back. Indeed, after performing the task, and before assuming the role of guardian, the robot exits the TAM moving forward and in a straight line for about 0.25 m. As a result, when the robot stops and takes the role of the guardian, the guarded task is behind its back.

The guardian can be in four different states, depending on the feedback received after performing the guarded task and on the state of chain construction.

- If the feedback is positive and the chain does not exist yet, the guardian is isolated in front of the first task. In this case, the guardian acts also as the tail of the chain to recruit chain members and form the first branch of the chain.

- If the feedback is negative and the chain is not perceived in the neighborhood, the guardian is isolated in front of either the second or third task. In this case, the guardian waits to be reached by the chain. In the meantime, it indicates that runners that have not performed any task should not perform the one it guards.

- If the feedback is negative and the branch of chain being built has arrived in the vicinity, the guardian joins the chain and completes the construction of the branch. If the construction of the chain is not complete, that is, if some tasks have not been reached yet, the guardian takes also the role of tail and initiates the construction of a further branch.

- Independently of the feedback originally received, after joining the chain and after being relieved from the duties of tail, the guardian limits itself to instruct the runners on the execution of the task guarded.

The transition between these four states is graphically described in Figure 3.5.

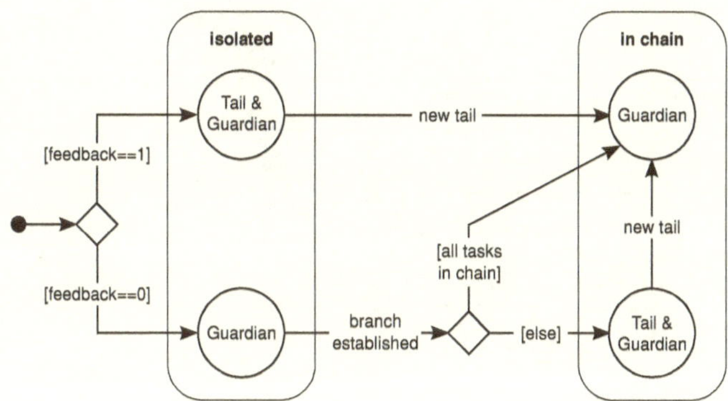

Figure 3.5: **State machine of a guardian.** The four states in which a guardian can be, with respect to the construction of the chain.

When acting as a tail, the guardian initiates the construction of a branch of chain on its right-hand side—the guarded task being behind its back. The goal is to construct a branch that departs from the right-hand side of the task and then sweeps the space anticlockwise when its tail reaches a wall. In all other respects, the guardian acts as any other tail—see Section 3.3-Tail.

The primary goal of a guardian is to guide runners in the execution of tasks. We will refer to the communication protocol between runners and guardians as *guardian protocol* (G protocol). The G protocol works as follows—see Figure 3.6 for an example of message exchange and Table 3.1 for a description of the messages. Every guardian locally broadcasts a range-and-bearing message (G_ADV) that contains information about the guarded task. The message contains two pieces of information: (i) the number CND of tasks that a runner must have performed to qualify for performing the guarded one; and (ii) the Boolean CONF, which states whether the guardian is confident about the value indicated in CND. CND and CONF are respectively encoded as DATA[3][2:0] and DATA[2][7]—see Figure 3.3.

As mentioned above, a runner that encounters an unattended task performs it and becomes its guardian. As a guardian, if the feedback received after performing the task was positive, it will broadcast CND = 0 and CONF = 1. This means that a runner that has not performed any task will be positively instructed to perform the guarded one. On the other hand, if the feedback was negative, the task is not the first to be performed but the information available is insufficient to tell whether the task is the second or the third of the sequence. In this case, the guardian will broadcast CND = 1

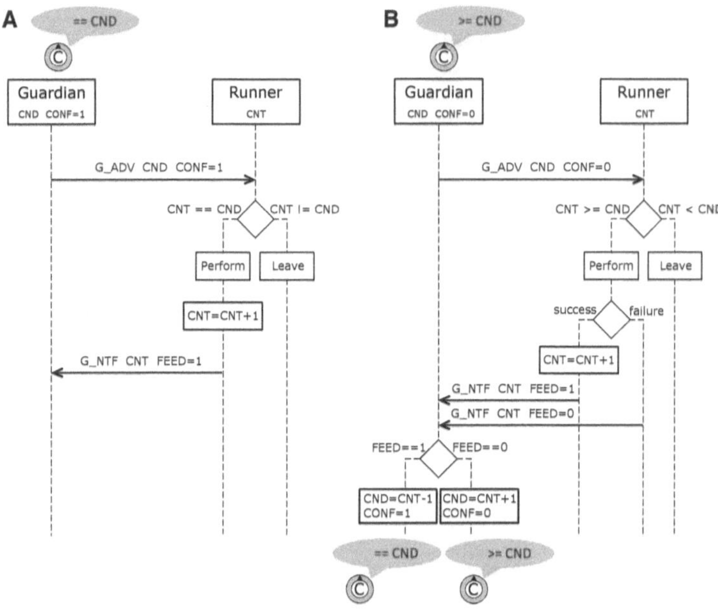

Figure 3.6: **Guardian protocol (G protocol), sequence diagram.** A runner that receives a G_ADV message performs the guarded task depending on whether the guardian is confident (CONF) about the value of CND. (**A**) If the guardian is confident about the value of CND, the runner performs the task provided that CNT = CND. (**B**) If the guardian is not confident about the value of CND, the runner performs the task provided that CNT >= CND. In the latter case, the guardian uses the feedback notified by the runner through a G_NTF message to update CND and CONF: If the runner has successfully performed the task in the correct order, the guardian sets CONF = 1 and CND = CNT − 1; otherwise, the guardian confirms CONF = 0 and sets CND = CNT + 1.

Table 3.1: **Guardian protocol (G protocol), description of messages.**

message	full name	description
G_ADV	advertisement message	Message locally broadcast by a guardian for nearby runners; it provides the conditions under which the task should be performed and the confidence of the guardian; it contains CND and CONF
G_NTF	notification message	Message sent by a runner to a guardian after executing its task; it notifies the guardian of the feedback received upon execution; it contains CNT and FEED (which is 1 if the execution was successful, 0 otherwise)

and CONF = 0. This means that a runner needs to have performed at least one task to qualify for performing the guarded task. Though, this condition is not sufficient for guaranteeing that the execution will be successful (CONF = 0).

The guardian is notified by every runner that performs the guarded task through a range-and-bearing message (G_NTF). Specifically, after performing the task, a runner communicates whether the execution was successful (DATA[1][1], see Figure 3.3) and the number CNT (DATA[3][2:0]) of tasks that it has so far successfully performed in the right order—including the guarded one, in case its execution was successful. If the guardian was not confident about the condition CND, that is, if CONF = 0, it updates its G_ADV as follows. If the runner has successfully performed the task in the correct order, the guardian sets CONF = 1 and CND = CNT − 1; otherwise, the guardian confirms CONF = 0 and sets CND = CNT + 1. The rationale is that, if the execution was unsuccessful, the runner failed to perform its task number CNT + 1 in the correct sequence. This means that the task guarded by the guardian is not the number CNT+1 in the sequence. At the same time, as the runner has already performed CNT tasks, the one guarded by the guardian is not among the first CNT of the sequence, otherwise the guardian would have been previously notified of a success and would have already set CONF. All in all, this means that the guarded task must be at least in position CNT + 2 of the correct sequence. As a consequence, the guardian will indicate that a runner must have performed at least CNT + 1 tasks before it qualifies for performing the one at hand.

A further duty of a guardian is to relay inhibition messages produced by the runners to avoid that they crowd around a task—see Section 3.3-Runner for the details.

Link

The chaining behavior we implemented for TS-Swarm is loosely inspired by previous work (Nouyan et al., 2008, 2009). Instead of the polychromatic cyclic pattern previously used to indicate the sense in which the chain should be followed, the links in TS-Swarm adopt a monochromatic pattern. This is because in TS-Swarm LED colors are used both to identify the different roles that the robot can assume, and to signal the presence of tasks. Additionally, e-pucks can display and recognize only a relatively limited variety of colors. Therefore, all the links indicate their role and mark their position in space by displaying the color yellow through their RGB LEDs. This means that runners move along the chain always in the same sense—that is, keeping the chain always on their left-hand side (see 3.3-Runner).

The goal of the chain is to assist the navigation of runners and ultimately lead them to the guardians of the tasks. To facilitate the navigation of runners, each link positions

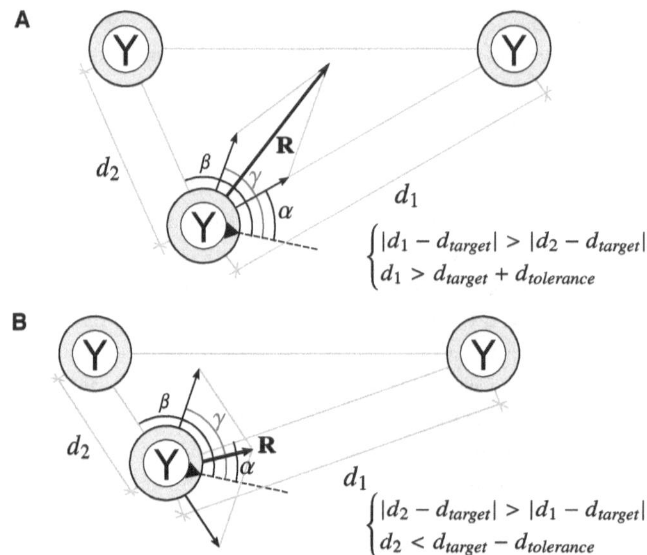

Figure 3.7: **Motion of a link.** Every movement of a link consists of two components: distance adjustment and alignment. The two components are computed with respect to the two closest neighboring chain members that the link perceives. The alignment component lies in the direction given by the mean (γ) of the angles under which the link perceives the following chain member (α), and the preceding one (β). Angles are measured with respect to the heading of the link, which is indicated by the black triangle and the dashed line. The distance adjustment component lies in the direction of the neighboring chain member whose distance is the farthest from the target spacing distance between chain members. (**A**) If the neighboring chain member is too far, the distance adjustment component is oriented towards it. (**B**) If the neighboring chain member is too close, the distance adjustment component is oriented in the opposite sense.

itself and aligns properly with respect to its two neighboring chain members: the one that precedes and the one that follows in the chain. At every control step, each link adjusts its position through displacements that consist of two components: distance adjustment and alignment (Figure 3.7).

The direction of the *distance adjustment* component is defined by the line connecting the link itself and the neighboring chain member that is the farthest away (in absolute value) from a target distance of 0.15 m. The distance adjustment component is oriented towards this neighboring chain member if the latter is farther that the target distance (Figure 3.7A). Otherwise, it is oriented in the opposite sense (Figure 3.7B).

The direction of the *alignment* component is given by the mean of the angles α

and β under which the preceding and following chain members are seen. Formally, the angle γ between the heading of a link and its alignment component is:

$$
\gamma = \begin{cases}
\arctan\left(\frac{\sin(\alpha)+\sin(\beta)}{\cos(\alpha)+\cos(\beta)}\right), & \text{if } \cos(\alpha)+\cos(\beta) > 0; \\[2ex]
\arctan\left(\frac{\sin(\alpha)+\sin(\beta)}{\cos(\alpha)+\cos(\beta)}\right) + \pi, & \text{if } \cos(\alpha)+\cos(\beta) < 0 \text{ and } \sin(\alpha)+\sin(\beta) \geq 0; \\[2ex]
\arctan\left(\frac{\sin(\alpha)+\sin(\beta)}{\cos(\alpha)+\cos(\beta)}\right) - \pi, & \text{if } \cos(\alpha)+\cos(\beta) < 0 \text{ and } \sin(\alpha)+\sin(\beta) < 0; \\[2ex]
+\frac{\pi}{2}, & \text{if } \cos(\alpha)+\cos(\beta) = 0 \text{ and } \sin(\alpha)+\sin(\beta) > 0; \\[2ex]
-\frac{\pi}{2}, & \text{if } \cos(\alpha)+\cos(\beta) = 0 \text{ and } \sin(\alpha)+\sin(\beta) < 0; \\[2ex]
\text{undefined}, & \text{if } \cos(\alpha)+\cos(\beta) = 0 \text{ and } \sin(\alpha)+\sin(\beta) = 0.
\end{cases}
$$

To avoid oscillations, a link does not adjust its distance or alignment if its position with respect to its neighboring chain members is sufficiently close to the ideal one. The tolerance on the distance is $\pm 0.05\,\mathrm{m}$ and the one on the alignment angle is $\pm 10°$.

A link stops in its current position, halting the adjustment of its alignment and spacing, upon being notified that the branch it composes is established. The notification (encoded in DATA[2][6]—see Figure 3.3) is sent by the guardian that has established the branch upon becoming the new tail of the chain; it is then relayed by the links.

Tail

The tail marks the position that the chain has reached and the point from which the construction should proceed. To announce its role and location, the tail uses a unique color for its RGB LEDs: magenta (Figure 3.8). In addition, the tail locally broadcasts, through its range-and-bearing board, a message that announces its role by setting DATA[1][2]—see Figure 3.3. The runners that navigate along the chain are thus informed when they approach the end of the chain and can attempt to join it to take the role of the new tail. Joining the chain is a critical process that must be robust to failure, promote the formation of a well-organized chain, and prevent inconsistent states. In particular, the process must guarantee that the chain has always one and only one tail. This is achieved via a communication protocol that regulates the interaction between the incumbent tail and the prospective one. The communication is performed via the range-and-bearing board. We call the protocol *tail protocol* (T protocol). The T protocol works as follows—see Figure 3.9 for an example of message exchange and Table 3.2 for a description of the messages.

The tail first makes sure to be at the right distance from the preceding chain mem-

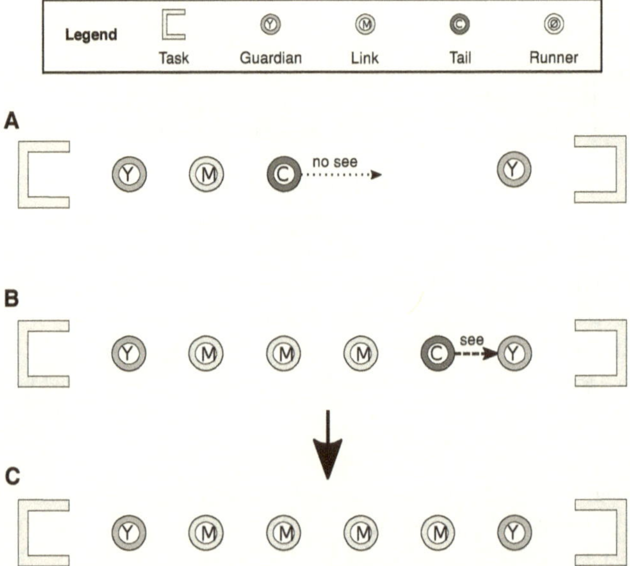

Figure 3.8: **Tail.** (**A**) The tail marks the position that the chain has reached at a given moment in time during its construction. It is identified by the color magenta. (**B**) When the tail sees a guardian in its vicinity, it halts the construction of the current branch of the chain. (**C**) The tail becomes a link and transfers its role of tail to the guardian.

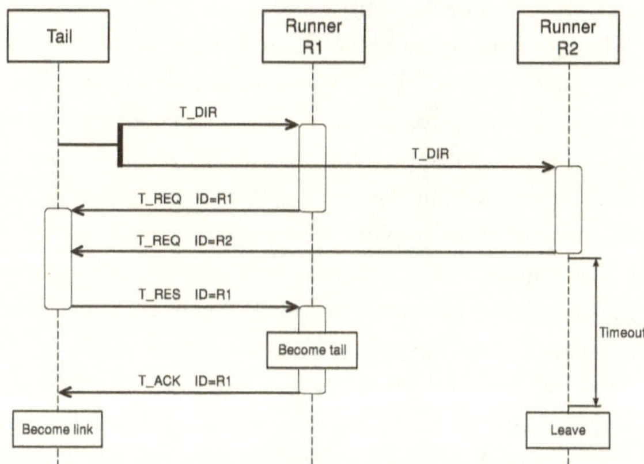

Figure 3.9: **Tail protocol (T protocol), sequence diagram.** The tail broadcasts a T_DIR message in the circular sector in which a new tail should join the chain. Both runners R1 and R2 receive this message and request to join as the new tail through a T_REQ. The T_REQ contains the ID of the runner that is sending the request. The tail selects the first request received, in the example, the one of runner R1. Then, the tail answers by sending a T_RES that contains the ID of the request selected, in this case, R1. After receiving the T_RES containing its own ID, R1 acknowledges the tail with a T_ACK message and becomes the new tail. When the tail receives the T_ACK of R1, it becomes a link. In the meantime, the request of R2 expires without a response and runner R2 leaves.

Table 3.2: **Tail protocol (T protocol), description of messages.**

message	full name	description
T_DIR	directional message	Message broadcast by a tail only in the circular sector in which a new tail should place itself; it indicates to a runner that it is in the right position to join the chain as the new tail
T_REQ	request message	Message sent by a runner to the tail; it represents the request to become the new tail; it contains the unique ID of the runner
T_RES	response message	Message sent by the tail to a runner that previously sent a request; it represents for the runner the confirmation of being the selected one to become the new tail; it contains the unique ID of the runner selected as the new tail
T_ACK	acknowledgment message	Message sent by the selected runner to the tail; it represents the acknowledgment of the response message and the acquisition of the role of tail; it contains the ID of the runner, which is now the tail

ber. Then, the tail starts broadcasting a directional message T_DIR (DATA[1][1], see Figure 3.3). Through this message, the tail indicates the circular sector in which the new tail should position itself. The ideal direction along which the new tail should position itself is computed by extending the segment connecting the tail with the preceding chain member: as a result, the new tail will be already correctly aligned. If a runner receives the directional message, it is because it is in the intended circular sector. In this case, the runner stops in place and sends the tail a request message T_REQ (DATA[1][0]). The request message carries the unique ID of the runner (DATA[0][7:0]). The tail might receive multiple requests from different runners. Requests are handled following a first-come-first-served policy. The tail responds with a T_RES (DATA[1][0]) that contains the ID of the runner selected as the new tail (DATA[0][7:0]). Upon reception of a response containing its own ID, the selected runner becomes the new tail and closes the communication via an acknowledgment message T_ACK (DATA[1][2] is set and its own ID is inserted in DATA[0][7:0]). The tail transfers its role to the runner after either receiving a T_ACK or detecting another robot that displays the magenta tail color.

Besides managing the communication with runners that attempt to join the chain, the tail is also in charge of searching for a guardian. When a guardian is found, the tail includes it in the branch of chain being built to complete its construction. To search for a guardian, the tail drives the branch of chain being built so that it sweeps the environment in an anticlockwise motion. The tail slowly moves perpendicularly to the direction along which the branch is being built, on the left-hand side of the latter (Figure 3.10C). As links react by adjusting spacing and alignment, the movement of the tail creates a domino effect: the whole branch turns pivoting around the guardian from which the branch originates, which stays still in its position in front of its task. As we have seen in Section 3.3, a branch departs from its originating guardian on its right-hand side—with the TAM on the back of the guardian (Figure 3.10B). During its construction, a branch reaches the wall of the arena, which prevents its further extension (Figure 3.10C). Thanks to the sweeping motion generated by the movement of the tail, the branch being built disentangles from the wall and allows further runners to join and extend it (Figure 3.10D).

To decide when it should start to move and trigger the sweeping motion of the branch, the tail cannot rely on its ability to detect the wall of the arena in its vicinity. This is due to the limited range of the proximity sensors and to the noise that affects their readings. As an alternative, the tail measures the amount of time during which no robot attempts to join the branch. Past a predefined threshold, the tail assumes that the branch being built is stuck against a wall and triggers the sweeping motion.

Once triggered, the sweeping motion is interleaved with pauses to allow links to align and, possibly, new members to join.

When the tail spots a guardian (Figure 3.10E), the sweeping motion stops and the tail aligns with the guardian and the preceding chain member. The guardian and the tail then establish the branch of the chain: the tail becomes a link and transfers its role of tail to the guardian (Figure 3.10F). The guardian then broadcasts a notification (DATA[2][6]—see Figure 3.3) for the links that compose the branch. Upon receiving this notification, the links halt the adjustment of their alignment and spacing, stopping in place. A new branch of the chain can now depart from the guardian. This mechanism eventually allows the swarm to form a chain that connects all the tasks, under the assumptions that the arena is convex, the tasks are located along its perimeter, and there are no obstacles in the environment (Figure 3.10).

Runner

Runners are all the robots that are not part of the chain. When the system is deployed, all the robots of the swarm start as runners. At runtime, they might assume the role of chain members—i.e., guardians, links, or tail—depending on the interactions they have with peers and environment. Runners do not adopt any color to announce their role. Indeed, their goal is not to landmark the space nor to provide information about task execution, but rather to follow the indications provided by the chain (Figure 3.11).

A runner might become a guardian if, while performing a random walk, it encounters an unattended task. This typically happens at the beginning of the lifetime of the system. The number of runners that become guardians equals the number of tasks to be performed.

The other runners continue to explore the environment by performing a random walk. If they encounter a chain, they follow it. Runners perceive the position of chain members via their omnivision module. Runners navigate along the chain by keeping it on their left. This generates an anticlockwise movement of the runners around the chain. The traffic of runners flows neatly along a branch of the chain in both senses without interfering: the runners that travel in one sense remain on one side of the chain, those that travel in the other sense remain on the other (Figure 3.12).

At every control step, a runner that follows a chain considers up to two neighboring chain members to define its direction of motion. The direction is given by the sum of a radial and a tangential component. The radial component keeps the robot at a target distance $\bar{d}$ from the chain while the tangential component allows the robot to proceed along it. The two components refer to a circle centered in a reference point. If the runner perceives only one chain member, the reference point is the position

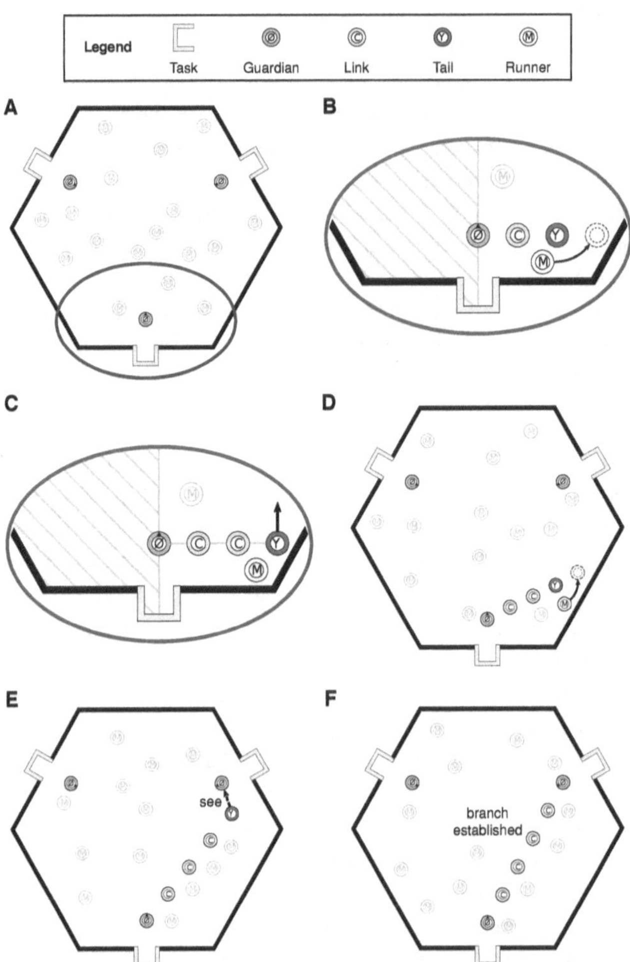

Figure 3.10: **Construction and motion of a branch of chain.** (**A**) Three robots become guardians of the three tasks. (**B**) The one that receives positive feedback after performing its guarded task initiates the construction of a branch of chain on its right-hand side. (**C**) The branch extends until it reaches the wall of the arena, which prevents other runners from joining it. The tail then moves perpendicularly to the direction of the branch, on the left-hand side of the latter. (**D**) The movement of the tail triggers a sweeping motion of the whole branch, which disentangles the branch from the wall and enables its further extension. (**E**) The process is repeated until the tail spots the guardian of a task. (**F**) The tail transfers its role to the guardian and completes the construction of the branch.

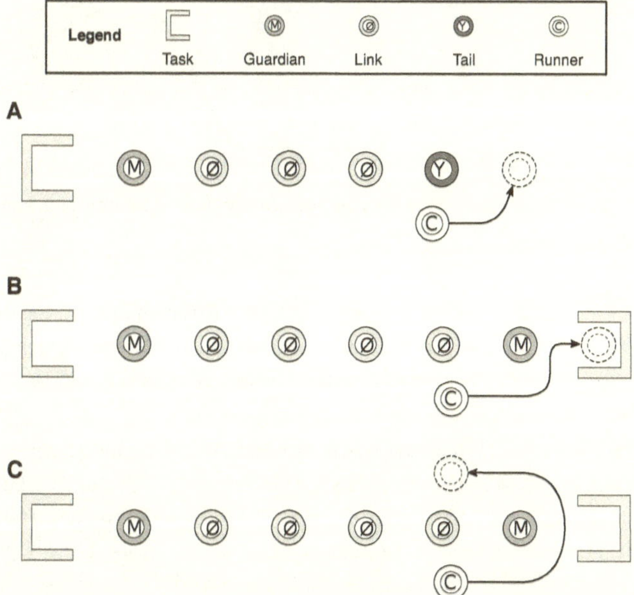

Figure 3.11: **Runners.** Runners are not identified by a color. (**A**) If a runner navigates along a branch of chain that has not been completed yet, it eventually encounters the tail and attempts to join the chain as the new tail. (**B**) If a runner navigates along a branch of chain that is complete, it eventually encounters a guardian. The guardian might indicate that the runner must (**B**) perform the task, or (**C**) skip it.

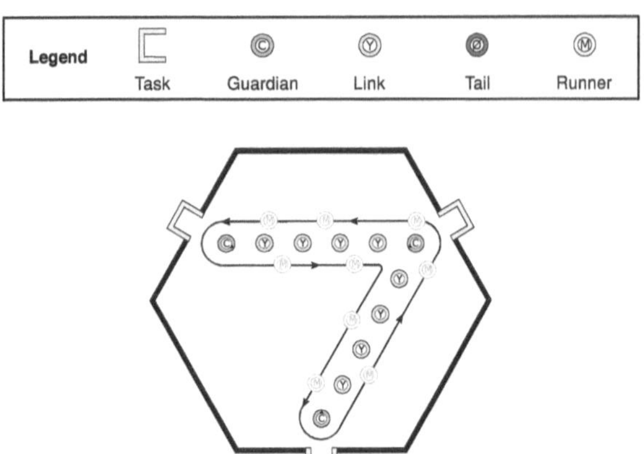

Figure 3.12: **Trajectory followed by the runners around the chain.** Runners navigate along the chain by keeping it on their left.

of the chain member perceived (Figure 3.13A,B). If the runner perceives two chain members, the reference point is the intersection between the line passing through the two chain members perceived and its perpendicular passing through the runner itself (Figure 3.13C,D).

Runners are led by the chain towards a guardian. When a runner reaches a guardian, it is instructed on whether to perform the guarded task—see Section 3.3-Guardian and Figure 3.6. In particular, upon receiving a G_ADV message, a runner evaluates the information provided in light of the number CNT of tasks it has already performed in the correct order. The runner performs the task either if (i) CONF = 1 and CND = CNT; or if (ii) CONF = 0 and CND <= CNT. Otherwise, the runner skips the task signaled by the guardian. After performing the task—or skipping it—the runner revolves anticlockwise around the guardian until it finds a branch of the chain to follow. If the task was performed, while revolving around it, the runner notifies the guardian of the feedback received upon task execution—see also Section 3.3. The runner notifies the guardian via a G_NTF in which it encodes its (possibly incremented) value of CNT as DATA[3][2:0] and the feedback received as DATA[1][1]—see Figure 3.3. Eventually, the chain leads each runner through all the tasks. By using the indication provided by the guardians, runners are able to perform the tasks along the chain in the correct order. When a runner completes the execution of a full sequence, it starts a new execution, in a cyclic manner.

To avoid overcrowding around the guardians and to enable an ordered execution of

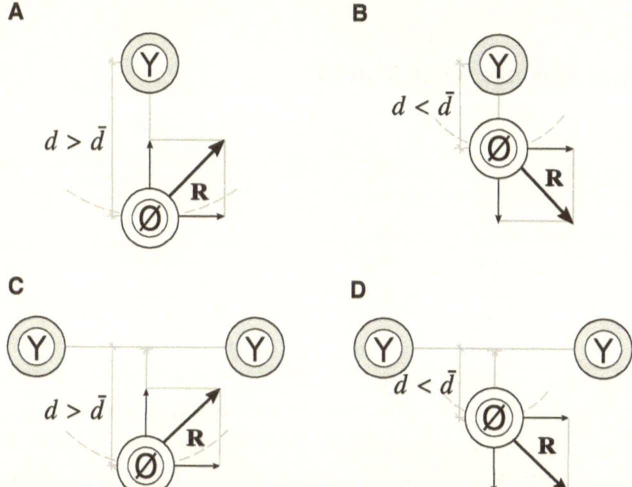

Figure 3.13: **Motion of a runner along a branch of the chain.** The direction of motion is given by the sum of a radial and a tangential component that refer to a circle (dashed gray line) centered in a reference point. (**A**, **B**) When only one chain member is perceived, the reference point is the position of that chain member. (**C**, **D**) When two chain members are perceived, the reference point is the intersection between the line passing through those chain members and its perpendicular passing through the runner itself. (**A**, **C**) If the runner is too far from the reference point ($d > \bar{d}$), the radial component is oriented towards it. (**B**, **D**) If it is too close ($d < \bar{d}$), the radial component is oriented in the opposite sense.

the tasks, runners adopt a mechanism of inhibition. When a runner starts approaching a task, it broadcasts an inhibition message through its range-and-bearing board. The inhibition message is echoed by the guardian of the task, if present. The nearby runners that perceive the inhibition message transition to an inhibition state or, with a certain probability, leave the area. Inhibited runners clear the way in front of the task to avoid hindering other robots' movement. After a certain period of time elapsed without receiving any inhibition message, an inhibited runner transitions back to normal operation.

3.4   Experiments with Mark I

In this section we describe the experiments that we performed to demonstrate the task-sequencing ability of Mark I, we analyze the results obtained, and we propose a set of possible improvements to the system.

3.4.1   Experimental design

The goal of the experiments we describe is to demonstrate Mark I and provide evidence that it is able to successfully sequence tasks in an autonomous and distributed way. First, we demonstrate Mark I_3 both in reality with a swarm of 20 e-puck robots and in simulation (Section 3.4.2). Besides showing the effectiveness of Mark I_3, this first set of experiments provides an assessment of the simulator (Section 3.4.3). After having shown that the simulator satisfactorily predicts the behavior of TS-Swarm on the e-puck robots, we adopt the simulator to perform a number of studies that we would be unable to perform with real robots. These studies either involve a large number of robots (more than those that we have available) or last longer than the battery life of the robots. In particular, we perform a study in which we assess the scalability of Mark I_3 by running experiments in which the number of robots ranges from 20 to 80 and the surface of the arena in which they operate ranges from $2.10\,\mathrm{m}^2$ to $33.67\,\mathrm{m}^2$ (Section 3.4.4). We also perform a study in which we assess the robustness of Mark I_3 to the number of robots comprised in the swarm (Section 3.4.5, and Table 3.4). We then perform a set of experiments to demonstrate the task-sequencing ability of Mark I_4, its scalability and its robustness (Section 3.4.6). Finally, in Section 3.5, we analyze the limitations of the currents implementation of Mark I and we propose some ideas for future improvements.

In these experiments we consider abstract tasks represented by TAMs (see Section 3.2.2). Robots operate in a bounded arena delimited by walls, which are $42\,\mathrm{mm}$ high. The arena is a regular hexagon when the tasks to be sequenced are three and a regular octagon when the tasks are four. The TAMs abstracting the tasks are distributed along the perimeter of the arena and are positioned in the middle of alternate sides. Each task is associated with a color. When the tasks are three, the colors are red (R), green (G), and blue (B). When they are four, the fourth color is orange (O). In each experimental run, the initial position of the robots, the correct sequence, and the relative position of the tasks are decided randomly. In all the supplementary movies (Garattoni and Birattari, 2020), for clarity, the correct sequence is always RGB when the tasks are three and RGBO when they are four.

In all experiments, the final goal of TS-Swarm is to perform the correct sequence of tasks ten times within a given time cap. As a performance measure, we consider the time required to complete one, five, and ten executions of the correct sequence. The first execution indicates that TS-Swarm has been able to solve the task-sequencing problem. The tenth execution determines the final success of the system and therefore the end of the experiment. The fifth execution represents the mid point of the previous two measures and provides visual information on whether the execution time grows linearly with the number of correct sequences performed or not.

Statistics

We report the performance of TS-Swarm via its empirical run-time distribution. Given: (a) one of the variants of TS-Swarm—i.e., Mark I_3, Mark I_4; (b) a specific experimental setting—e.g., a setting characterized by the number of robots, the surface of the arena, and the time cap; and (c) a target objective—i.e., the execution of one, five, or ten correct sequences, we perform k independent runs and we observe, for each run, the time required to attain the target objective. The empirical run-time distribution is the empirical distribution of these observations.

Formally, let TC be the time cap of each run, $j \in \{1,\ldots,k\}$ be the index of a run, r_j be the run-time of run j, and $k' \leq k$ be the number of successful runs, that is, those runs $j : r_j < $ TC. The empirical run-time distribution is defined as $RTD(t) \triangleq \hat{P}_s(\tau \leq t) = \#\{j \mid r_j \leq t\}/k$. Here, $\hat{P}_s(\tau \leq t)$ is an estimate of the probability that the system attains its target objective in an amount of time τ that is less than or equal to t. In other words, the empirical distribution $RTD(t) \triangleq \hat{P}_s(\tau \leq t)$ is an estimate of the probability of success of the system over time (up to TC). For a given target objective and in a given experimental setting, the success ratio of the system within the time cap TC is $S_{\text{TC}} \triangleq k'/k$.

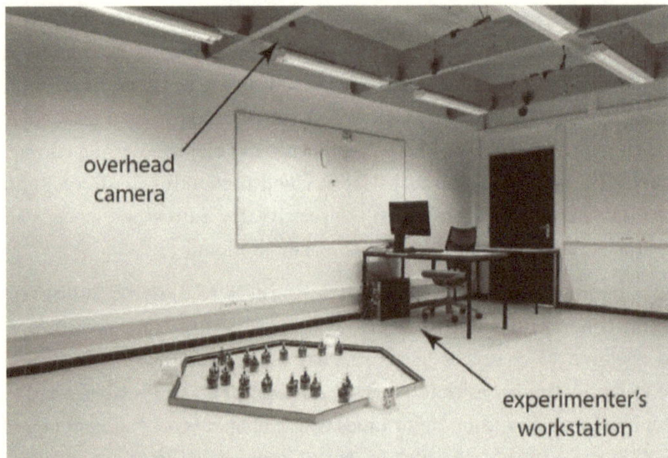

Figure 3.14: **Experimental setting.** The hexagonal arena, 3 TAMs, and 20 e-pucks. The setting includes also an overhead camera for filming experiments and a workstation for coordinating the TAMs and gathering data.

3.4.2 Robot experiments

We run Mark I_3 10 times with 20 e-pucks. The experiments are performed in a controlled environment with a flat surface and uniform light conditions. The arena where the robots operate is a regular hexagon with sides of 0.9 m. A camera operating at about 3 frames per second is mounted on the ceiling with its axis lying on the vertical line passing through the center of the arena. The experimental setting is pictured in Figure 3.14.

We present the results of 10 consecutive runs. The performance of Mark I_3 in each of these 10 runs concurs to the statistics presented: no observed result is discarded for any reason whatsoever. Once a run starts, it is accounted for in the statistics. The statistics include therefore also the failures. In Table 3.3, we report the lab notebook, which includes the record of all the information that we collected during each of the 10 runs. A run is terminated either at the tenth execution of the correct sequence or at a time cap of 40 minutes (2,400 s). A typical run of Mark I_3 on the robots is displayed in movie S1 of the online supplementary material Garattoni and Birattari (2020). All the supplementary movies feature overlays to illustrate the different robot roles and clarify how TS-Swarm solves the task-sequencing problem. The complete results of the robot experiments are reported in Figure 3.17A.

Concerning the experiments with the robots, Mark I_3 was able to complete ten sequences before the time cap on 9 runs out of 10. In the remaining run, it was

Table 3.3: **Laboratory notebook.** Results of 10 consecutive runs of Mark I_3 on 20 e-puck robots operating in an hexagonal area of $2.10\,\mathrm{m}^2$. Record of all the information that we collected during each of the 10 runs.

run	date	begin	end	e-puck failures	time to completion of				notes
					chain	1 seq	5 seq	10 seq	
1	Oct 11, 2016	1:26 pm	1:45 pm	Runner fails at 370 s without causing problems	290 s	410 s	758 s	1,198 s	Chain functional and built quickly; smooth execution of 10 sequences
2	Oct 11, 2016	3:57 pm	4:23 pm	Runner fails near the chain at 1,120 s without causing problems	565 s	599 s	916 s	1,506 s	Chain functional but too spaced near the last guardian; this slows the execution
3	Oct 12, 2016	11:43 am	12:06 pm	No failures	490 s	590 s	890 s	1,250 s	Chain highly functional thanks to the two branches rather open; smooth execution of 10 sequences
4	Oct 12, 2016	2:40 pm	3:18 pm	No failures	595 s	909 s	1,420 s	2,248 s	Communication problems in chain construction, but construction completes; slow execution due to a guardian too far from TAM (blue)
5	Oct 13, 2016	11:09 am	11:27 am	No failures	302 s	340 s	680 s	1,070 s	Chain straight and functional; smooth execution of 10 sequences
6	Oct 13, 2016	2:48 pm	3:24 pm	No failures	350 s	495 s	772 s	1,410 s	Chain overcrowded but functional, first branch not perfectly aligned; smooth execution of 10 sequences
7	Oct 14, 2016	11:13 am	11:33 am	Runner fails at 600 s without causing problems	320 s	362 s	564 s	923 s	Chain well aligned and functional; smooth execution of 10 sequences
8	Oct 14, 2016	2:40 pm	3:06 pm	Runner fails at 600 s near a TAM hindering other runners. Communication problems in the construction of the first branch. Runners have problems to enter the blue TAM	425 s	455 s	1,125 s	1,600 s	Communication problems in the construction of the first branch; a runner fails and hinders other runners slowing the execution
9	Oct 18, 2016	10:28 am	11:08 am	Runner fails at 1,850 s without causing problems	605 s	988 s	1,553 s	time cap	Chain overcrowded and poorly aligned near the guardians; this slows the execution; at time cap, 9 sequences performed
10	Oct 18, 2016	1:26 pm	1:54 pm	Two runners fail (at 530 s and 1,460 s) without causing problems	319 s	411 s	810 s	1,656 s	Chain functional despite the first branch is sparse; smooth execution of 10 sequences

nonetheless able to determine the correct order of the tasks and to perform 9 sequences before the time cap. In the laboratory notebook (Table 3.3), we report the record of all the information we collected during the 10 runs with the robots. Movie S1 of the online supplementary material (Garattoni and Birattari, 2020) shows a typical run on Mark I_3 with the robots.

3.4.3   Assessment of the simulator

Alongside the experiments with the robots, we perform similar experiments in simulation using ARGoS, with the idea of producing an assessment of the simulation environment. The control software used in the two sets of experiments is the same: after performing the robot experiments, we port the control software back to the simulated environment without any modification. Because performing experiments in the simulated environment is much less time consuming than performing them in reality, we gather results on 30 simulated runs. Moreover, because battery life is not a concern in simulation, we extended the duration of runs beyond the time cap of 40 minutes. Results are reported in Figure 3.17B, while a set of overhead snapshots of robot experiments is shown in Figure 3.15.

Concerning the simulated experiments, Mark I_3 achieved a success ratio of 90%: in 27 runs out of 30, Mark I_3 was able to complete ten sequences before the time cap. Figure 3.16A shows an overhead snapshot of a simulated experiment with Mark I_3. On its right, a QR that links to the online video.

All in all, the results show that Mark I_3 is able to sequence and perform the three given tasks both on the robots and in simulation. Moreover, they indicate that the simulation environment we adopted allows us to predict the performance of the robots. Movie S2 in the online supplementary material (Garattoni and Birattari, 2020) shows a typical run in simulation, side by side with one on the robots.

3.4.4   Scalability study

We perform simulated experiments in five experimental settings. In each setting, we double the surface of the arena with respect to the previous one. We also increase the number of robots by a factor of $\sqrt{2}$. The rationale is that, by increasing the surface of the arena by a factor 2, the distance between the TAMs increases by a factor $\sqrt{2}$. We therefore expect the number of robots that become chain members to grow roughly by the same factor. By increasing the swarm size by a factor $\sqrt{2}$, we expect that the robots will be sufficiently many to connect all the TAMs. The control software adopted in the scalability study is exactly the same in all the settings. The parameters that characterize the five settings are given in Table 3.4. We ran Mark I_3 30 times in each

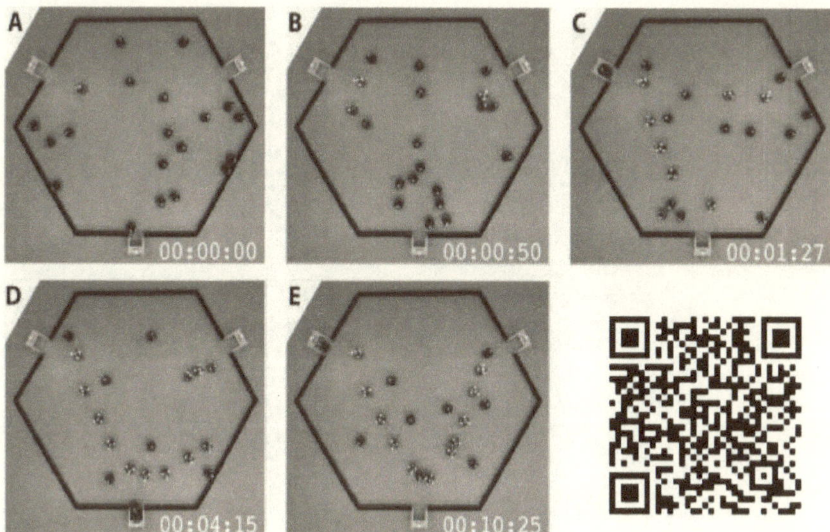

Figure 3.15: **Overhead snapshots of robot experiments.** (**A-E**) Mark I_3, robot experiments (Movie S1). Supplementary movies can be found in the online supplementary material (Garattoni and Birattari, 2020). To watch the video, either click on the QR code at the bottom right or scan it with the camera of a mobile device.

of the five settings (Figure 3.18A). Movie S3 of the supplementary material (Garattoni and Birattari, 2020) shows the highlights of the scalability study.

In Figure 3.19(A to E), we present the results of the scalability study by reporting the empirical run-time distributions for each setting. They comprise curves for the successful execution of one, five, and ten correct sequences. The results show that Mark I_3 scales well with the size of the arena and the number of robots. They also confirm our hypothesis that by increasing the number of robots by the same factor of the arena's side, the robots are sufficiently many to connect all the TAMs. This is made explicit by the plot presented in Figure 3.20.

3.4.5 Robustness study

We use the same five experimental settings considered in the scalability study. For each of the five settings, we vary the number of robots with respect to the one adopted in the scalability study. We consider both a smaller (-10% and -20%) and a larger number of robots ($+20\%$, $+40\%$, $+60\%$, $+80\%$, and $+100\%$). For each experimental setting and each number of robots tested, we report the run time distribution for the

Figure 3.16: **Overhead snapshots, Mark I$_m$ in simulation.** (A) Mark I$_3$, simulation (Movie S2, side-by-side with a run on the robots). (B) Mark I$_4$, simulation (Movie S4). Supplementary movies can be found in the online supplementary material (Garattoni and Birattari, 2020). To watch one of the videos, either click on the corresponding QR code or scan it with the camera of a mobile device.

successful execution of ten sequences and the empirical distribution of the number of robots in the chain. We ran Mark I$_3$ 30 times for each number of robots considered in each of the five settings (Table 3.4 and Figure 3.18A).

In Figure 3.19(F to J), we present the results of the robustness study in which we analyze the robustness of Mark I$_3$ to the variation of the number of robots w.r.t. the default. The run-time distributions in Figure 3.19(F to J) confirm that Mark I$_3$ is in general robust to the variation of the number of robots. Increasing slightly the number of robots is even beneficial to the performance in all the settings, while decreasing it quickly degrades the performance, as there are no longer enough robots to reliably connect the three TAMs. The distributions of the number of chain members as a function of the total number of robots—reported in Figure 3.19(K to O)—show that, given the size of the arena, Mark I$_3$ places roughly the same number of robots to connect the three TAMs, independently of the total number of robots that the swarm comprises.

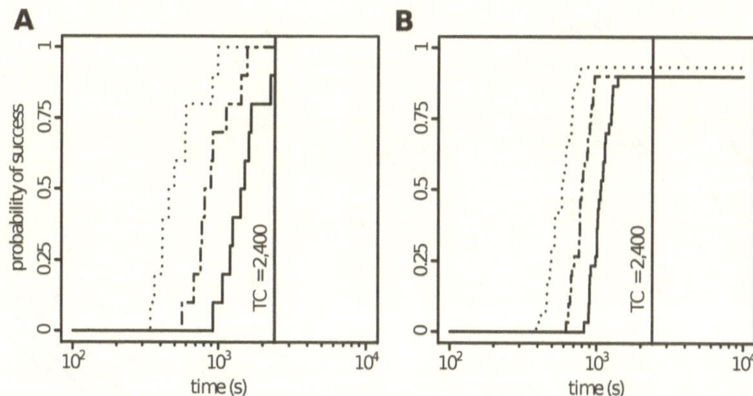

Figure 3.17: **Empirical assessments of Mark I$_3$.** Empirical run-time distributions for the execution of 1 (dotted lines), 5 (dot-dash lines), and 10 (solid lines) sequences. (**A**) Mark I$_3$, robot experiments. (**B**) Mark I$_3$, simulation.

Table 3.4: **Parameters of the scalability and robustness studies on Mark I.** We report the parameters that characterize each experimental setting in which Mark I$_m$ is studied. The scalability study is performed using the default number of robots in each setting. Between one setting and the following one, we double the surface of the arena in which the robots operate. The robustness study is performed varying the number of robots between -20% and +100% with respect to the default number of each setting.

	setting	arena's side	arena's area	number of robots								time cap
				-20%	-10%	**default**	+20%	+40%	+60%	+80%	+100%	
Mark I$_3$	0	0.90 m	2.10 m^2	16	18	**20**	24	28	32	36	40	2,400 s
	1	1.27 m	4.21 m^2	23	25	**28**	34	39	45	50	56	3,400 s
	2	1.80 m	8.42 m^2	32	36	**40**	48	56	64	72	80	4,800 s
	3	2.55 m	16.84 m^2	45	51	**57**	68	80	91	103	114	6,800 s
	4	3.60 m	33.67 m^2	64	72	**80**	96	112	128	144	160	9,600 s
Mark I$_4$	0	0.66 m	2.10 m^2	18	20	**22**	26	31	35	40	44	100,000 s
	1	0.93 m	4.21 m^2	25	28	**31**	37	43	50	56	62	100,000 s
	2	1.32 m	8.42 m^2	35	40	**44**	53	62	70	79	88	100,000 s
	3	1.87 m	16.84 m^2	50	56	**62**	74	87	99	112	124	100,000 s
	4	2.64 m	33.67 m^2	70	79	**88**	106	123	141	158	176	100,000 s

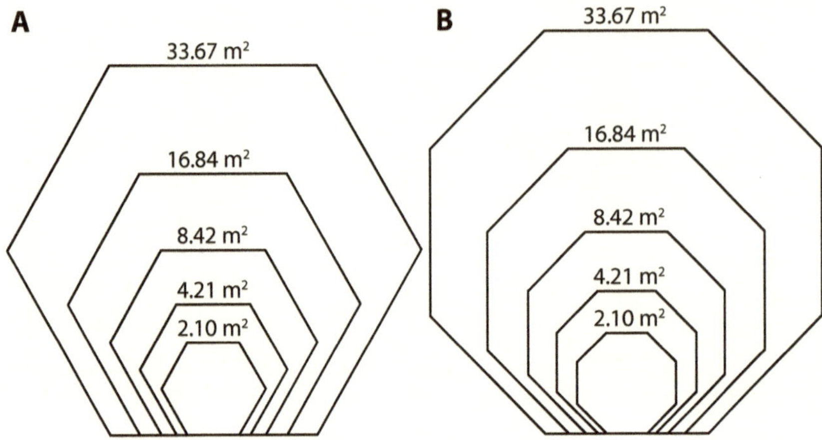

Figure 3.18: **Scalability and robustness analysis, the arenas.** Shape and size of the arenas considered for the scalability and robustness study of (**A**) Mark I_3 and (**B**) Mark I_4.

3.4.6   Experiments with Mark I_4

To complete a sequence, a robot must perform four tasks in a specific order, which is *a priori* unknown. The arena is a regular octagon with sides of 0.66 m. To connect the four TAMs, Mark I_4 needs to establish three branches of chain: one more than the two that Mark I_3 needs to establish. For this reason, we consider here a swarm of 22 robots, rather than 20 as we did for Mark I_3. We also increase the time cap to 100,000 s (i.e., about 28 h). We run Mark I_4 30 times in simulation. A typical run is displayed in movie S4 of the supplementary material (Garattoni and Birattari, 2020). Figure 3.16B shows an overhead snapshot of a simulated experiment with Mark I_4. On its right, a QR that links to the online video. Finally, we study the scalability and the robustness of Mark I_4 (summary of experimental settings in Table 3.4 and Figure 3.18B). Results are reported in Figure 3.21 and Figure 3.22.

The results in Figure 3.21 indicate that TS-Swarm can successfully sequence and perform more that three tasks without being subject to substantial modifications.

As it should have been expected, it takes slightly longer to sequence four task than three, but the success ratio is the same obtained by Mark I_3: 27/30.

The scalability study reported in Figure 3.22(A to E) indicates that Mark I_4 scales well with the size of the arena and the number of robots. Figure 3.22(F to J) confirm that Mark I_4 is also robust to the variation of the number of robots. Slight increases of the number of robots are mostly beneficial to the performance in all the settings,

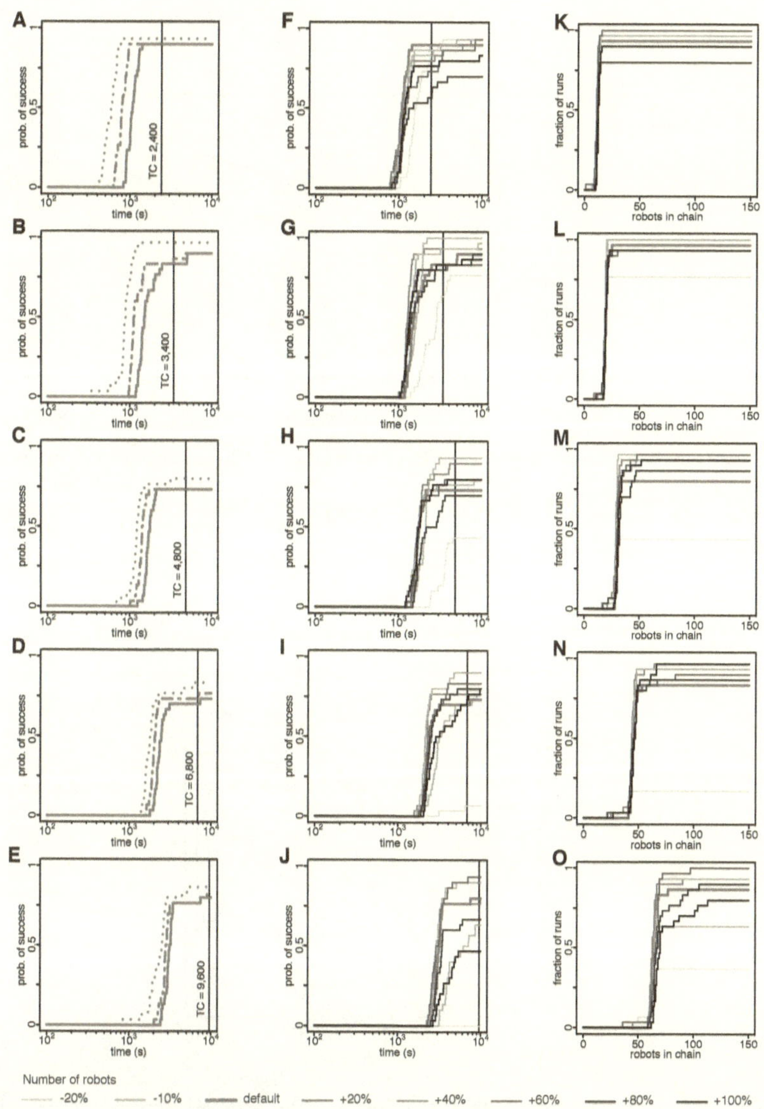

Figure 3.19: **Scalability and robustness of Mark I₃.** (**A-E**) Scalability study using the default number of robots in five arenas of different size (see Table 3.4). Empirical run-time distribution for the execution of one (dotted), five (dot-dash), and ten (solid) sequences. (**F-J**) Robustness to the variation of the number of robots between -20% and +100% of the default number (see Table 3.4). Empirical run-time distribution for the execution of ten sequences. (**K-O**) Empirical distribution of the number of robots in the chain as a function of the total number of robots. Arena's area: (**A, F, K**) $2.10\,\mathrm{m}^2$. (**B, G, L**) $4.21\,\mathrm{m}^2$. (**C, H, M**) $8.42\,\mathrm{m}^2$. (**D, I, N**) $16.84\,\mathrm{m}^2$. (**E, J, O**) $33.67\,\mathrm{m}^2$.

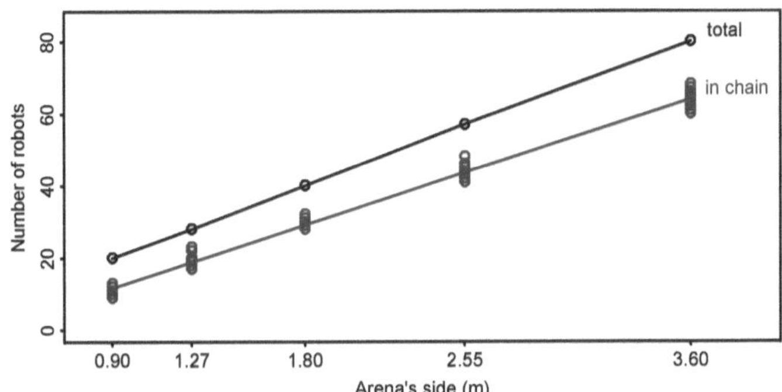

Figure 3.20: **Number of chain members in Mark I$_3$.** Chain members in the scalability study presented in Figure 3.19. The plot confirms that the number of chain members grows linearly with the side of the arena, as we assumed.

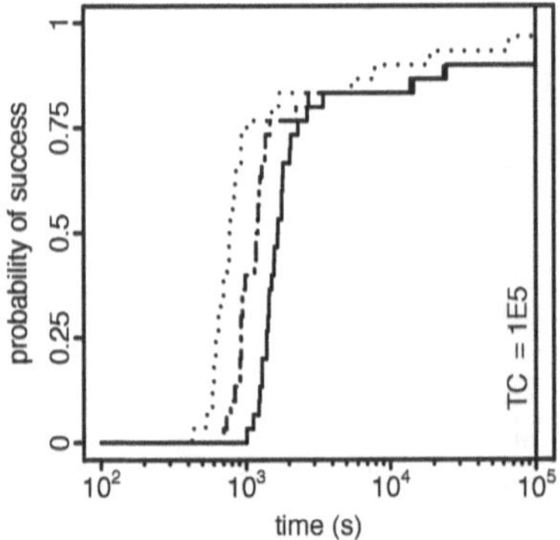

Figure 3.21: **Empirical assessment of Mark I$_4$.** Empirical run-time distributions for the execution of 1 (dotted lines), 5 (dot-dash lines), and 10 (solid lines) sequences. Mark I$_4$, simulation.

whereas decreases degrade the performance due to a lack of a sufficient number of robots to reliably connect the four TAMs. Differently from Mark I_3, drastically increasing the number of robots w.r.t. the default also produces a lower performance. This is probably due to the fact that, when the density is too high, robots might interfere with each other or hinder each other's movements. The distributions of the number of robots in chain as a function of the total number of robots—reported in Figure 3.22(K to O)—confirm the observations made for Mark I_3: given the size of the arena, Mark I_4 uses roughly the same number of robots to connect the four TAMs, independently of the total number of robots that the swarm comprises. In all the settings, for at least 50% of the runs the number of chain members is independent of the total number of robots—with the exception of the -20% and -10% cases in which the number of robots is apparently insufficient for establishing a chain reliably.

3.5 Limitations and possible improvements

In the following, we list and describe a series of limitations in our implementation of Mark I. For each limitation we also propose a possible future improvement.

Transmission of robot IDs limits scalability. The scalability of Mark I is limited by the fact that robots include their identifier in the range-and-bearing messages they broadcast (see Figure 3.3). For this **book**, we ran experiments with up to 216 robots. A single byte is therefore sufficient to encode a globally-unique identifier for each robot. Indeed, with one byte, up to 256 robots can be uniquely identified. With two bytes, we would be able to uniquely identify up to 65,536 robots. Should the number of robots be even larger, globally-unique identifiers could be too demanding on the communication.

Possible improvement: as a possible alternative, we could adopt locally-unique identifiers, that is, identifiers that are unique within the range of communication of each robot. Locally-unique identifiers have been successfully demonstrated with a swarm of one thousand robots (Rubenstein et al., 2014b).

The number of tasks must be known at design time. Mark I_m assumes that the number of tasks to be sequenced (m) is known at design time. This limits the autonomy of the system, as the designer must input or hard code the value of m in the robot's behavior.

Possible improvement: we could let the swarm determine the number of tasks autonomously at run time. A possible solution would be to let the swarm continue to build branches of chain until the chain forms a closed loop. At that point, the number of tasks would just be equal to the number of branches (which is maintained during the

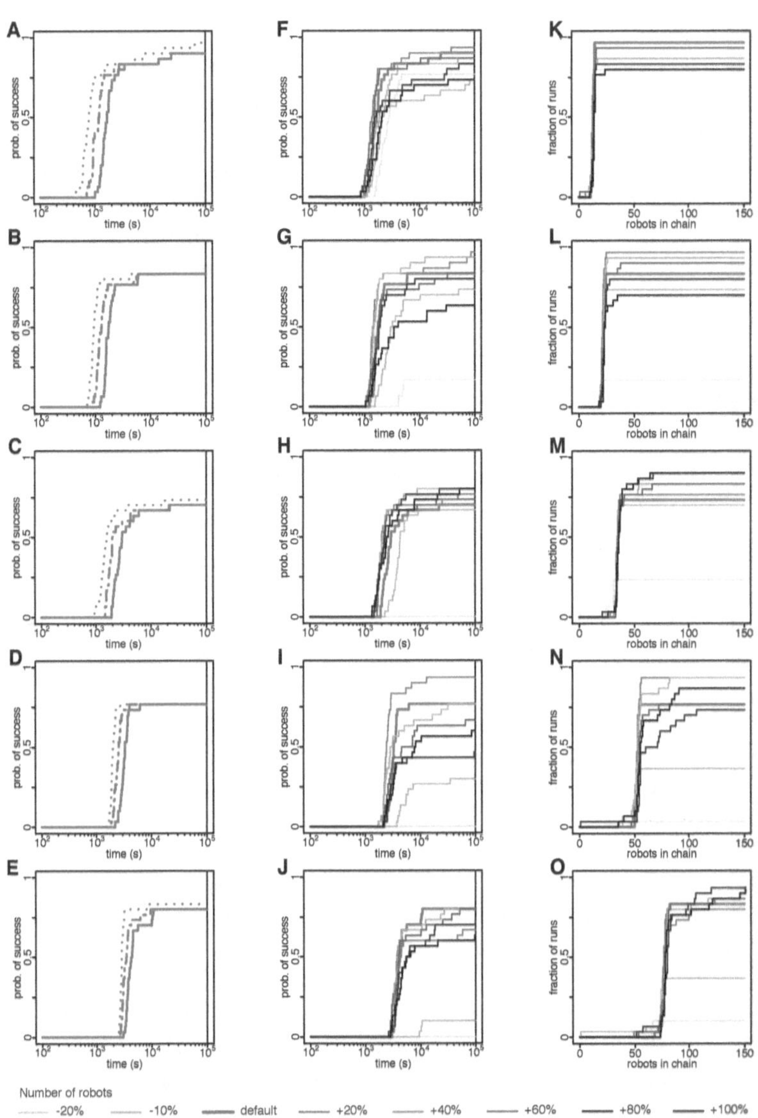

Figure 3.22: **Scalability and robustness of Mark I$_4$.** (**A-E**) Scalability study using the default number of robots in five arenas of different size (see Table 3.4). Empirical run-time distribution for the execution of one (dotted), five (dot-dash), and ten (solid) sequences. (**F-J**) Robustness to the variation of the number of robots between -20% and +100% of the default number (see Table 3.4). Empirical run-time distribution for the execution of ten sequences. (**K-O**) Empirical distribution of the number of robots in the chain as a function of the total number of robots. Arena's area: (**A, F, K**) 2.10 m^2. (**B, G, L**) 4.21 m^2. (**C, H, M**) 8.42 m^2. (**D, I, N**) 16.84 m^2. (**E, J, O**) 33.67 m^2.

construction of the chain), and this value could be propagated to all the robots along the chain.

Chains might overstep guardians. In some cases, the branch of chain being built fails to locate the guardian it is supposed to reach. This may be due to several factors, including (i) the lack of a sufficient number of robots to extend the branch up to the guardian; (ii) a temporary fault in the omnivision module of the tail. These factors cause the branch to overstep the guardian and eventually reach another branch of chain or the guardian from which the branch itself originates. As a result, the branch being built merges with another branch or collapses on itself. In both cases, the system fails.

Possible improvement: the tail could implement a mechanism to detect whether it is approaching another branch of chain, or the originating guardian of its own branch. Should this happen, the tail could invert the sense in which the branch sweeps around its originating guardian. By alternating clockwise and anticlockwise sweeps, the branch being built would explore the environment more effectively and increase the chance to spot the guardian it is supposed to reach.

Robots' movement is unsophisticated. Chain members and runners move in a simple and unsophisticated way. For simplicity, we implemented the links so that they stop moving upon being notified that the branch they form is established. If a runner bumps into a link pushing and it away from the correct position, the continuity of the chain is broken and the functionality of the whole system is compromised. This is more likely to occur when the swarm is comprised by a large number of robots and the arena is large. In these cases, the branches are long and need to be functional for a long time.

Possible improvement: More refined movement mechanisms could be implemented in order to make the motion of chain members and runners more precise and reliable. Links could keep adjusting alignment and spacing even after their branch is established. In particular, they could benefit from a mechanism to regain their original position, should they detect that the functionality of their branch is compromised.

Chaining rests on restrictive assumptions. The chaining behavior works under the assumptions that (i) the arena is convex, (ii) the tasks are located along its perimeter, and (iii) there are no obstacles.

Possible improvement: we could relax these assumptions if the path between guardians were obtained by first covering the space with a lattice-like formation, and then selecting the shortest path on this lattice. Robots that are not on the shortest path could leave; those that are on the shortest path would remain to act as way-points. An approach to select the shortest path on a lattice has been demonstrated with a swarm of e-pucks (Campo et al., 2010). This approach is based on artificial pheromone.

Single points of failure, the guardians.   In the current implementation of Mark I_m, guardians fullfil a crucial role for the system, as they signal the presence of tasks and instruct runners to execute the tasks in the right order. If one of them fails, the system is not able to continue the execution and thus the task-sequencing mission is not completed.

Possible improvement: we could leverage the redundancy of the robot swarm to recover from guardian failures. The chain members in the vicinities of guardians could actuate mechanisms of error recovery. For instance, the chain member that precedes a guardian could detect the failure and replace the guardian.

Chapter 4

TS-Swarm Mark II

The version of TS-Swarm presented in the previous chapter (Mark I) relies on a strong assumption: runners receive feedback after performing each single task. Under this assumption, a runner is notified that the sequence being performed is wrong as soon as it performs the first task that departs from the correct sequence. In this chapter, we remove this assumption by considering the case in which a runner has to perform an entire sequence before receiving any feedback. We present and describe TS-Swarm Mark II, hereafter Mark II, a robot swarm that can solve this new, more challenging, task-sequencing problem.

4.1 Description of Mark II

The lack of feedback immediately after performing a task makes the sequencing problem harder. This new task-sequencing problem is a combinatorial problem whose computational complexity is $O(m!)$, where m is the number of tasks. Moreover, the immediate feedback received after the execution of their task allowed the guardians of Mark I to break the symmetry among themselves and thus start the construction of the chain from a single point. This is not possible in Mark II, as guardians do not receive any feedback after performing their respective task.

In the following of this section, we describe the modifications that we made to Mark I to produce Mark II, a robot swarm that can sequence tasks in the case in which robots have to perform an entire sequence before knowing whether the sequence itself is correct or not. The platforms that we used to develop Mark II are the same used for Mark I—i.e., e-puck robots, TAMs to abstract the execution of tasks, and ARGoS as development environment and robotic simulator. A description of the platforms is given in Section 3.2 and in Appendix A.

4.1.1   Mark II$_3$

TS-Swarm Mark II$_3$, hereafter Mark II$_3$, assumes that a runner has to perform an entire sequence of three tasks before knowing whether the sequence itself is correct or not. Here, we only describe the differences between Mark I$_3$ and Mark II$_3$, as in all other respects, the two systems are identical and thus we refer the reader to Chapter 3 for the other implementation details. Mark II$_3$ differs from Mark I$_3$ in the following main points:

- All guardians initiate the construction of a branch of the chain in parallel. When all the branches are complete, the chain is a closed loop that connects all the guardians. The guardians use the closed-loop chain to exchange information and to establish an initial sequence of the tasks.

- Once the initial sequence is established, the guardians direct the runners so that the initial sequence and all its permutations are tested, one after the other. The swarm collectively explores the tree of the permutations of the initial sequence. Each robot (guardians, links, and runners) contributes to the collective exploration by acting reactively, relying only on partial knowledge of the sequences being tested.

- After completing a sequence, a runner receives feedback on whether the sequence is correct or not. It notifies the feedback to the guardian of the last task, which then shares it with all other guardians via the closed-loop chain. If the sequence is correct, from then on all runners are instructed so as to perform it. Otherwise, the following sequence in the depth-first search is tested.

All guardians initiate the construction of a branch because, after executing a task for the first time, they do not have any way to break the symmetry among themselves. Indeed, in Mark I$_3$ (and Mark I$_4$), after executing their respective tasks, one of the guardians receives positive feedback while the others a negative one. The one that receives the positive feedback is the one associated with the first task to be performed in the sequence. This is the guardian that initiates the construction of the chain. In Mark II$_3$, no guardian receives any feedback so the above mechanism cannot be used to break the symmetry at this moment of the system's lifetime. As an alternative, all guardians initiate the construction of a branch. When all branches are complete and the chain is closed (Figure 4.1A), the guardians communicate via range-and-bearing messages that are relayed by the chain members. The guardians use the communication medium that they have established to break the symmetry and order themselves. The order of the guardians (and therefore of the tasks they guard) defines the initial sequence to be tested and eventually the whole permutation tree that will be then searched.

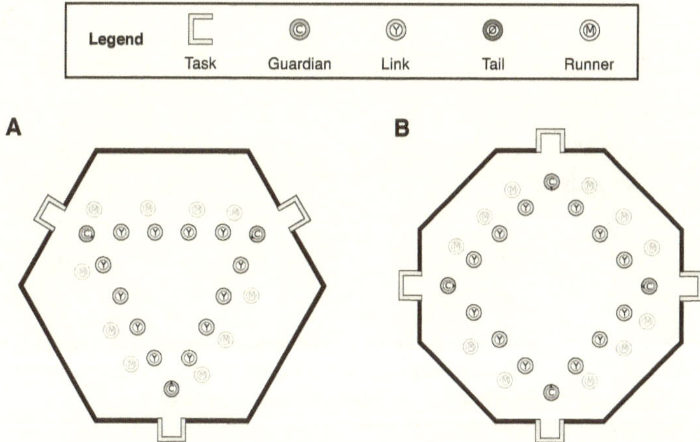

Figure 4.1: **The chain in Mark II$_3$ and Mark II$_4$.** The guardians do not have any way to break the symmetry among themselves after performing their guarded task. Therefore, they all initiate the construction of a branch of the chain. When all branches are complete, the chain forms a closed loop that the guardians use for communicating. (**A**) The chain built by Mark II$_3$; and (**B**) the one built by Mark II$_4$.

The guardians order themselves using a leader-election algorithm (Chang and Roberts, 1979) through which they assign the label c to the one of them with the largest ID. Guardian c sends a message that is relayed clockwise along the closed-loop chain. The first guardian that receives the message is assigned the label b and the following the label a.

Mark II$_3$ explores the space of the possible sequences by traversing the tree of the permutations of $\langle a, b, c \rangle$ (Figure 4.2). The tree is explored via depth-first search. In practice, the guardians coordinate themselves to direct the runners so that they test the possible sequences one after the other. As a first step, the guardians address the runners to the tasks guarded by a, b, and c, in this order; as a second step, to the tasks guarded by a, c, and b; as a third step to the tasks guarded by b, a, and c, and so on. The exploration of the permutation tree is distributed. At any moment in time, each guardian has only partial knowledge about the sequence being tested. To clarify this, consider the example in Figure 4.3, in which the guardian standing in front of the green task is assigned the label b via the election mechanism described above.

The transition between a step and the following one is triggered by a failure reported by a runner. When a runner that has performed all the three tasks in the order prescribed at that step receives negative feedback, it reports the failure to the guardian

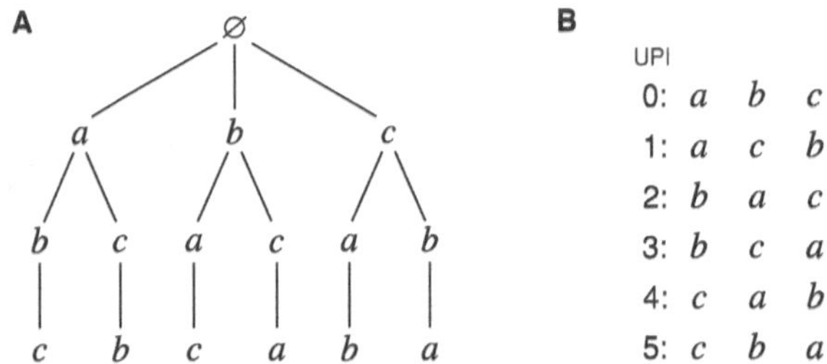

Figure 4.2: **Exploration of the space of possible sequences in Mark II$_3$. (A)** Permutation tree. **(B)** List of sequences explored via a depth-first search of the permutation tree. UPI is the *unique permutation identifier*: the sequence with UPI = 0 is the initial sequence defined via the election algorithm. UPI = 1 is its first permutation encountered by searching the permutation tree via depth-first search; UPI = 2 is the second one, and so on.

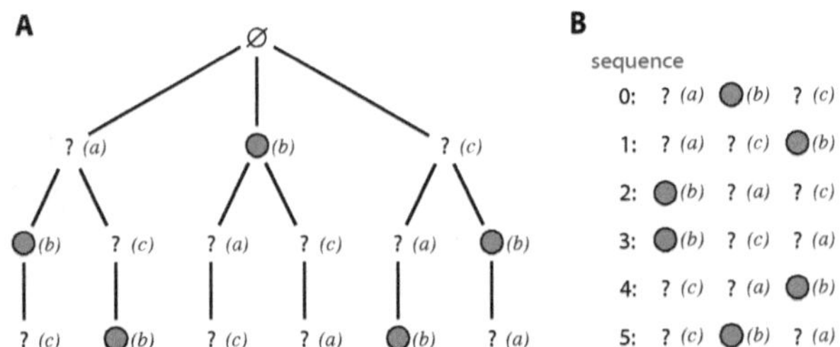

Figure 4.3: **Exploration of the sequence space in Mark II$_3$, as seen by the guardian of the green task.** In this example, the green task is the second of the initial sequence. Its guardian ignores the colors of the first and last tasks; it only knows that its label is b, and therefore, its task is second. More precisely, this guardian is in the state in which it directs to its task the runners that have performed exactly one task. This guardian (as the others) directs the runners throughout the search process without knowing the sequence that is tested at each step. At the first step, its task is second—it directs to its task the runners that have performed exactly one task. At the following step, its task is third—it directs to its task the runners that have performed exactly two tasks. Then, its task is first—it directs to its task runners that have not yet performed any task and so on. **(A)** Permutation tree generated by the guardian of the green task on the basis of its partial knowledge of the initial sequence. **(B)** Sequences explored through depth-first search of the permutation tree.

of the last task performed. This guardian initiates the transition to the following step by sending a range-and-bearing message that is relayed along the closed-loop chain. The guardians that receive the message transition to the following step. The exploration of the permutation tree is thus reactive: guardians transition from one sequence to the following one in response to the notification of a failure from a runner or from the closed-loop chain.

As they approach a guardian, the runners are notified of the transition and start from scratch the execution of the new sequence, as directed by the guardians.

Eventually, the search of the permutation tree leads the swarm to hit the correct sequence. When the first runner receives positive feedback after performing the three tasks in the correct order, it reports the success to the guardian of the last task performed. This guardian communicates the event to the other guardians via the closed-loop chain. The search process terminates and the guardians continue directing the runners following the policy that produced the correct sequence.

To summarize: in Mark II_3, the search process relies on the cooperation of guardians, links, and runners. Each robot acts reactively and no robot has complete knowledge of the sequence being performed: (i) runners navigate the environment following the chain and perform tasks if directed to do so by the guardians; (ii) links relay information; (iii) guardians transition from state to state on the basis of the feedback reported by the runners. In this context, the state of a guardian is "direct to the guarded task runners that have already performed X tasks". The sequence of states through which each guardian transitions is uniquely determined by its position in the initial sequence, as determined by the election process. At no moment in time, a guardian has knowledge of the whole sequence being performed by the runners. Also in the absence of immediate feedback, TS-Swarm collectively converges to the correct sequence without relying on representations of the environment or symbolic reasoning. The ability to sequence the given tasks emerges from the interaction of individual robots that act reactively.

Implementation details. The range-and-bearing message that the guardians exchange to transition from a step of the search to the following one contains a 2 bit field M4PI, *modulo 4 permutation identifier*. M4PI is an integer ranging from 0 to 3. Its value is M4PI = UPI mod 4, where UPI is the *unique permutation identifier*, of the new permutation to be tested (Figure 4.2B). The general scheme of the range-and-bearing message encoding in Mark II_3 is given in Figure 4.4.

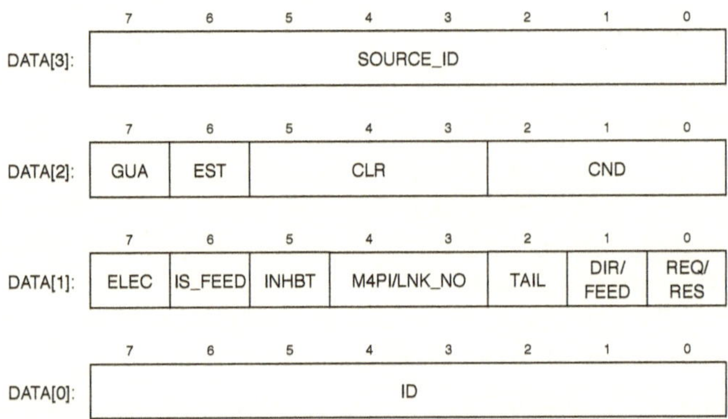

Figure 4.4: **Encoding of the range-and-bearing message in Mark II$_3$.** Guardians identify their messages by setting DATA[2][7]. When the closed-loop chain is complete, the guardians use a leader-election algorithm to break the symmetry among themselves. In the messages exchanged during the phase of leader election, the guardians set DATA[1][7]. Once a first order among themselves has been established, the guardians start to test the possible sequences of the tasks. By setting DATA[1][4:3], each guardian broadcasts locally the counter M4PI (modulo 4 permutation identifier). Its value is M4PI = UPI mod 4, where UPI is the unique permutation identifier associated with the permutation of the initial sequence that is currently being tested—see Figure 4.2. Each guardian broadcasts locally also the number CND of tasks that a runner must have performed to qualify for performing the guarded one. This is done by setting DATA[2][2:0]. Runners notify the guardians of the success/failure of the sequence being tested by setting DATA[1][6] and inserting the feedback received in DATA[1][1]. If the feedback is negative, the guardian that receives the notification switches to the next sequence and sends a message to the other guardians, which in turn complete the transition to the next sequence. In all the messages that they send along the closed-loop chain, guardians insert their ID in both DATA[0][7:0] (ID) and DATA[3][7:0] (SOURCE_ID). This enables a guardian that receives a message from the chain to identify the source guardian that sent it by reading the field SOURCE_ID. The field ID is instead overwritten by every link that relays the message along the chain. The fields that are not described here are equivalent to the ones used in Mark I$_3$ (and Mark I$_4$)—see Figure 3.3.

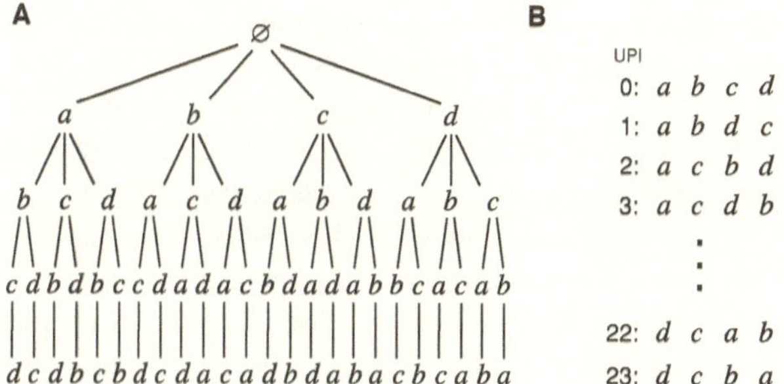

Figure 4.5: **Exploration of the space of possible sequences in Mark II$_4$.** (**A**) Permutation tree. (**B**) List of sequences determined by exploring the permutation tree via depth-first search. UPI is defined in the caption of Figure 4.2.

4.1.2 Mark II$_4$

TS-Swarm Mark II$_4$, hereafter Mark II$_4$, sequences four tasks under the assumption that a runner has to perform an entire sequence before knowing whether the sequence itself is correct or not. Mark II$_4$ is based on Mark II$_3$, the only difference being that the task counter of the robots counts up to four. In all other respects, the two systems are identical: the implementations, the values of all the parameters, and the encoding of the range-and-bearing messages (Figure 4.4) are the same.

In Mark II$_4$, four robots take the role of guardian, one per task, and the chain is a closed loop that comprises four branches (Figure 4.1B). The tree of permutations and the list of sequences explored by Mark II$_4$ is given in Figure 4.5.

4.2 Experiments with Mark II

In this section, we describe the experiments that we performed to demonstrate the task-sequencing ability of Mark II. The problem solved by Mark II is combinatorial and more complex ($O(m!)$) than that solved by Mark I, as runners have to complete a full sequence of tasks before being notified of whether the order of execution is correct or not.

As for Mark I, we analyze the results obtained in each set of experiments and finally we propose a set of possible improvements to the system.

4.2.1 Experimental design

The experimental design we adopted for Mark II is similar to the one used for Mark I.

Having shown that the simulator successfully predicts the behavior of TS-Swarm on e-puck robots (see Section 3.4.3), we perform a number of experiments in simulation to demonstrate Mark II. In particular, we perform a study in which we assess the scalability of Mark II$_3$ by running experiments in which the number of robots ranges from 25 to 100 and the surface of the arena in which they operate ranges from $2.10\,\mathrm{m}^2$ to $33.67\,\mathrm{m}^2$. We also perform a study in which we assess the robustness of Mark II$_3$ to the number of robots comprised in the swarm (Section 4.2.2, and Table 4.1). We then perform a set of experiments to demonstrate that also Mark II can sequence a greater number of tasks. We show the task-sequencing ability of Mark II$_4$, its scalability and its robustness (Section 4.2.3). Finally, in Section 4.3, we analyze the limitations of the currents implementation of Mark II and we propose some ideas for future improvements. In these studies, each run of the system lasts 100,000 s (i.e., about 28 h), which is much longer than the battery life of the e-puck robot. These studies would thus be impossible to perform on real e-pucks.

Similarly to the settings we used with Mark I, in all these experiments we abstract tasks using TAMs (see Section 3.2.2). Robots operate in a bounded arena delimited by walls, which are 42 mm high. The arena has the shape of a regular hexagon when the tasks to be sequenced are three and a regular octagon when the tasks are four. The TAMs are positioned in the middle of alternate sides of the arena. Each task is associated with a color. When the tasks are three, the colors are red (R), green (G), and blue (B). When they are four, the fourth color is orange (O). In each experimental run, the initial position of the robots, the correct sequence, and the assignment tasks-TAMs are decided randomly. In all the supplementary movies Garattoni and Birattari (2020), for clarity, the correct sequence is always RGB when the tasks are three and RGBO when they are four.

The final goal of TS-Swarm is to perform the correct sequence of tasks ten times within a given time cap. As a performance measure, we consider the time required to complete one, five, and ten executions of the correct sequence. We report the performance of the system via its empirical run-time distribution (see Section 3.4.1).

4.2.2   Experiments with Mark II$_3$

We consider a scenario in which a robot needs to complete an entire sequence of tasks before being notified of a possible error. The tasks to be sequenced are three. The correct sequence is *a priori* unknown. The arena is the same of the experiments with Mark I$_3$: a regular hexagon with sides of 0.9 m. As Mark II$_3$ needs to build a closed-loop chain, the number of robots it requires is larger than the one required by Mark I$_3$. We consider here a swarm of 25 robots rather than the 20 of the experiments performed

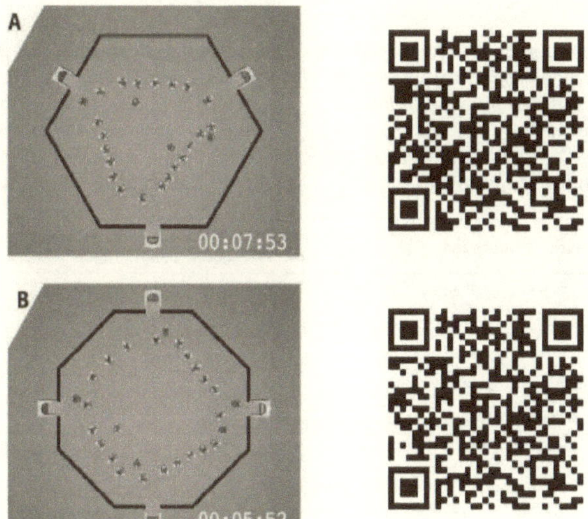

Figure 4.6: **Overhead snapshots, Mark II$_m$ in simulation.** (**A**) Mark II$_3$, simulation (Movie S5). (**B**) Mark II$_4$, simulation (Movie S6). Supplementary movies can be found in the online supplementary material (Garattoni and Birattari, 2020). To watch one of the videos, either click on the corresponding QR code or scan it with the camera of a mobile device.

with Mark I$_3$. As Mark II$_3$ must explore a relatively large space of solutions, we increase the time cap to 100,000 s. We run Mark II$_3$ 30 times in simulation. A typical run is displayed in movie S5 of the online supplementary material Garattoni and Birattari (2020). Figure 4.6A shows an overhead snapshot of a simulated experiment with Mark II$_3$. On its right, a QR that links to the online video. Finally, we study the scalability and the robustness of Mark II$_3$ (Table 4.1 and Figure 3.18). Results are reported in Figure 4.7A and Figure 4.8.

The performance of Mark II$_3$ is only slightly lower than the one of Mark I$_3$ but remains above the 75% line. In 24 out of the 30 runs, Mark II$_3$ successfully performs ten correct sequences within the time cap. This shows that TS-Swarm can cope with a sequencing problem in which the feedback is received only at the end of a sequence of tasks. The slightly lower success ratio with respect to Mark I$_3$ is possibly due to the challenge of building three branches of chain in parallel. Indeed, when the branches are built in parallel they risk to interfere or even merge with each other. The scalability study—reported in Figure 4.8(A to E)—indicates that Mark II$_3$ scales less well than Mark I$_m$, although the success ratio remains close to 50% even in the largest setting. A

Table 4.1: **Parameters of the scalability and robustness studies on Mark II.** We report the parameters that characterize each experimental setting in which Mark II$_m$ is studied. The scalability study is performed using the default number of robots in each setting. Between one setting and the following one, we double the surface of the arena in which the robots operate. The robustness study is performed varying the number of robots between -20% and +100% with respect to the default number of each setting.

	setting	arena's side	arena's area	number of robots								time cap
				-20%	-10%	**default**	+20%	+40%	+60%	+80%	+100%	
Mark II$_3$	0	0.90 m	2.10 m^2	20	22	**25**	30	35	40	45	50	100,000 s
	1	1.27 m	4.21 m^2	28	31	**35**	42	49	56	63	70	100,000 s
	2	1.80 m	8.42 m^2	40	45	**50**	60	70	80	90	100	100,000 s
	3	2.55 m	16.84 m^2	56	63	**70**	84	98	112	126	140	100,000 s
	4	3.60 m	33.67 m^2	80	90	**100**	120	140	160	180	200	100,000 s
Mark II$_4$	0	0.66 m	2.10 m^2	22	24	**27**	32	38	43	49	54	100,000 s
	1	0.93 m	4.21 m^2	30	34	**38**	46	53	61	68	76	100,000 s
	2	1.32 m	8.42 m^2	43	49	**54**	65	76	86	97	108	100,000 s
	3	1.87 m	16.84 m^2	61	68	**76**	91	106	122	137	152	100,000 s
	4	2.64 m	33.67 m^2	86	97	**108**	130	151	173	194	216	100,000 s

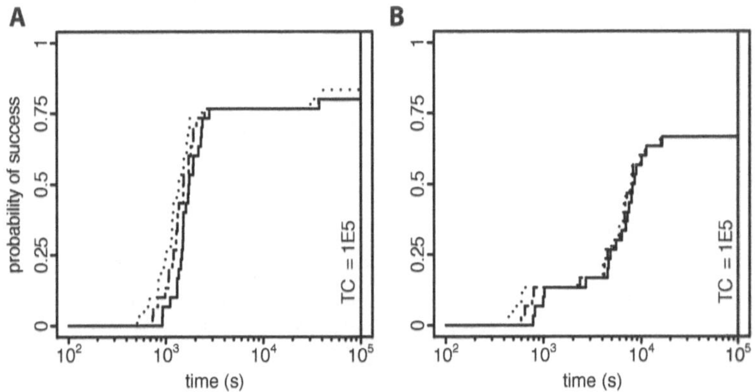

Figure 4.7: **Empirical assessment of Mark II.** Empirical run-time distributions for the execution of 1 (dotted lines), 5 (dot-dash lines), and 10 (solid lines) sequences. (**A**) Mark II$_3$, robot experiments. (**B**) Mark II$_4$, simulation.

factor that might impact negatively the performance of Mark II_3 is that links no longer adjust their position after their branch is established—see Section **??** for a detailed description of the robot control software and Section 4.3 for possible improvements. As a result, if a runner bumps into a link and pushes it away from the ideal position, the chain might be interrupted and the system might become unable to complete its mission. This is more likely to happen when the swarm comprises many robots and the arena is large. The possible improvement discussed under the heading "Robots' movement is unsophisticated" could contribute to increase the scalability of Mark II_3. The robustness study—reported in Figure 4.8(F to J)—indicates that, compared to Mark I_3, Mark II_3 is also less robust to variations of the number of robots, again possibly because these variations affect the chain construction process. Also in this case, we expect that the improvements to the chain construction and robots' movement that we propose could affect positively the performance of Mark II_3 in large arenas, when the number of robots departs from an ideal value. The distributions of the number of chain members as a function of the total number of robots—reported in Figure 4.8(K to O)—indicate that, in the successful runs, the number of robots that are used to connect the three TAMs is not affected by the total number of robots.

4.2.3 Experiments with Mark II_4

We consider a scenario similar to the one considered for Mark II_3, with the only difference that the tasks to be sequenced are four. The arena is the same of the experiments with Mark I_4: a regular octagon with sides of 0.66 m. We consider here a swarm of 27 robots: more than those considered in the experiments with Mark I_4 because Mark II_4 needs to build a closed-loop chain. Also in this case, the time cap is at 100,000 s. We run Mark II_4 30 times in simulation. A typical run is shown is movie S6 Garattoni and Birattari (2020). Figure 4.6B shows an overhead snapshot of a simulated experiment with Mark II_4. On its right, a QR that links to the online video. Finally, we study the scalability and the robustness of Mark II_4 (Table 4.1 and Figure 3.18). Results are reported in Figure 4.7B and Figure 4.11.

As it should have been expected, the success ratio of Mark II_4 is lower than the one of Mark II_3. Nonetheless, in 20 out of the 30 runs, Mark II_4 successfully sequences the four given tasks and performs ten correct sequences within the time cap. A comparison of the run-time distributions also shows that the time required by Mark II_4 to perform ten sequences of four tasks is longer than the one required by Mark II_3 to perform ten sequences of three tasks. Indeed, in 50% of the runs Mark II_3 performs ten correct sequences in less than 2,000 s. On the other hand, Mark II_4 reaches the 50% mark after 8,000 s.

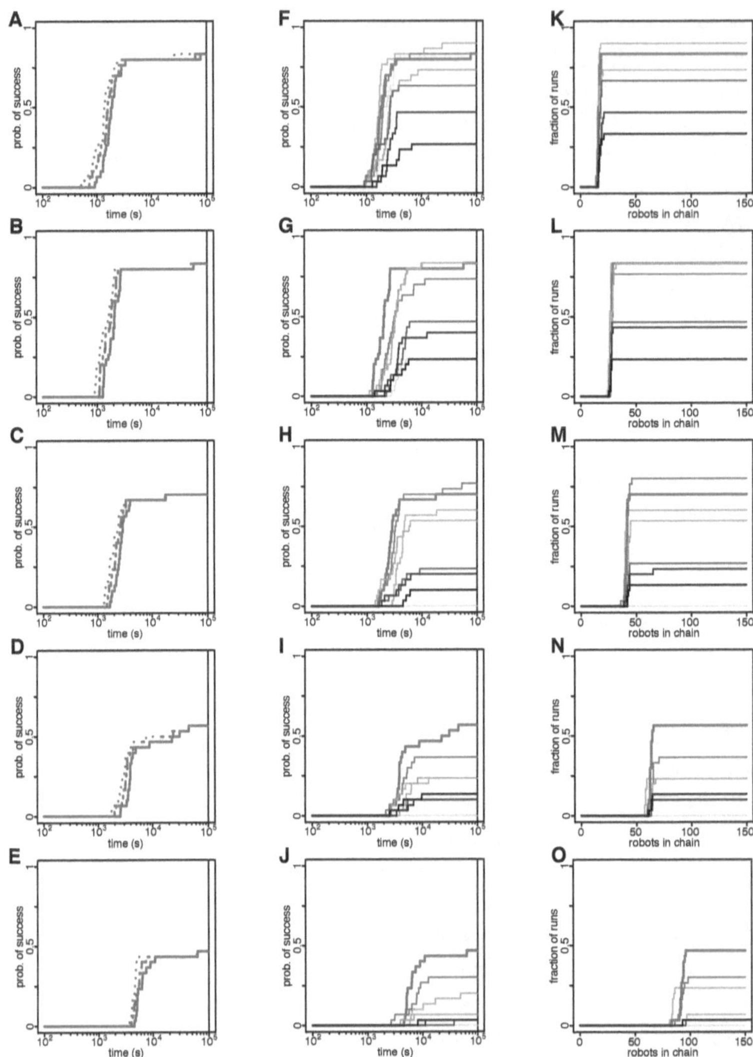

Figure 4.8: **Scalability and robustness of Mark II₃.** (**A-E**) Scalability study using the default number of robots in five arenas of different size (see Table 3.4). Empirical run-time distribution for the execution of one (dotted), five (dot-dash), and ten (solid) sequences. (**F-J**) Robustness to the variation of the number of robots between -20% and +100% of the default number (see Table 3.4). Empirical run-time distribution for the execution of ten sequences. (**K-O**) Empirical distribution of the number of robots in the chain as a function of the total number of robots. Arena's area: (**A, F, K**) 2.10 m². (**B, G, L**) 4.21 m². (**C, H, M**) 8.42 m². (**D, I, N**) 16.84 m². (**E, J, O**) 33.67 m².

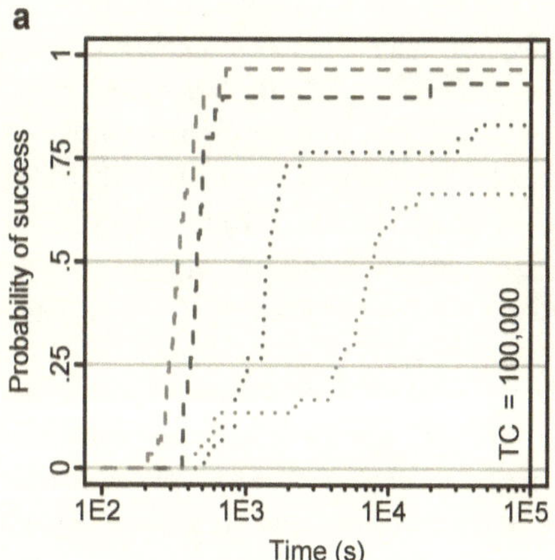

Figure 4.9: **Comparison between Mark II$_3$ (blue) and Mark II$_4$ (red).** Empirical runtime distributions for the construction of the chain (dashed line) and the successful execution of the first sequence (dotted line).

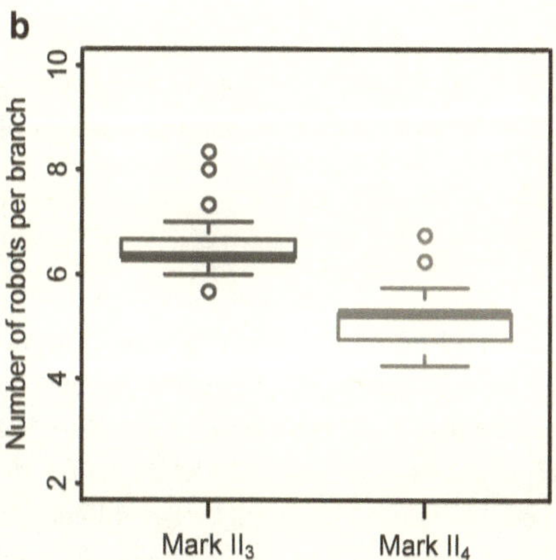

Figure 4.10: **Comparison between Mark II$_3$ (blue) and Mark II$_4$ (red).** Number of robots comprised in each branch of the chain.

Figure 4.9 shows that Mark II_3 and Mark II_4 require roughly the same time to construct the closed-loop chain. Actually, Mark II_4 is slightly faster. This is explained by the fact that in Mark II_4, the number of robots comprised in each branch of the chain is lower than in Mark II_3 (Figure 4.10). On the other hand, the time needed by Mark II_4 to perform the first sequence—that is, to solve the sequencing problem—is larger than the one needed by Mark II_3. Two factors contribute to this difference: (i) it takes longer to a runner to test a sequence in Mark II_4 than in Mark II_3 (4 tasks to perform vs. 3); and (ii) the space of the possible sequences explored by Mark II_4 is larger than the one explored by Mark II_3 (4! = 24 sequences searched by Mark II_4 vs. 3! = 6 searched by Mark II_3). The longer duration of a run in Mark II_4 with respect to Mark II_3—in particular, the longer time needed to explore the space of the possible sequences—is likely the reason why Mark II_4 achieves a lower success ratio than Mark II_3. Indeed, the longer the time needed to explore the space of the sequences, the longer the system (particularly the chain) needs to remain functional, and eventually, the higher the chance that something goes wrong. For example, the runners might push the links and/or the guardians out of position thus breaking the continuity of the chain—see Section 4.3 for a description of possible improvements.

The results shown in Figure 4.11 support the observations drawn so far. The challenges of constructing m branches of chain in parallel prevent Mark II_m from achieving a success ratio similar to the one of Mark I_m. It also lowers its scalability and robustness properties—Figure 4.11(A to E) and Figure 4.11(F to J). However, when Mark II_m completes the chain and solves the task-sequencing problem (execution of the first sequence), it reliably executes ten correct sequences. When this happens, the number of chain members is independent of the total number of robots in the swarm—Figure 4.11(K to O).

4.3   Possible improvements

In the following, we list and describe a series of limitation in our implementation of Mark II. Some limitations originate directly from the implementation of Mark I. In fact, all the limitations that are discussed in Section 3.5 for Mark I are inherited by Mark II. Some of them are repeated here because Mark II is affected by them in a different way or because the possible improvement suggested differs from the one discussed for Mark I. Other limitations are specific to the implementation of Mark II.

Transmission of robot IDs limits scalability.   As in Mark I, the scalability of Mark II is limited by the fact that robots include their identifier in the range-and-bearing messages they broadcast (see Figure 4.4). Also, in Mark II guardians use their identifier in

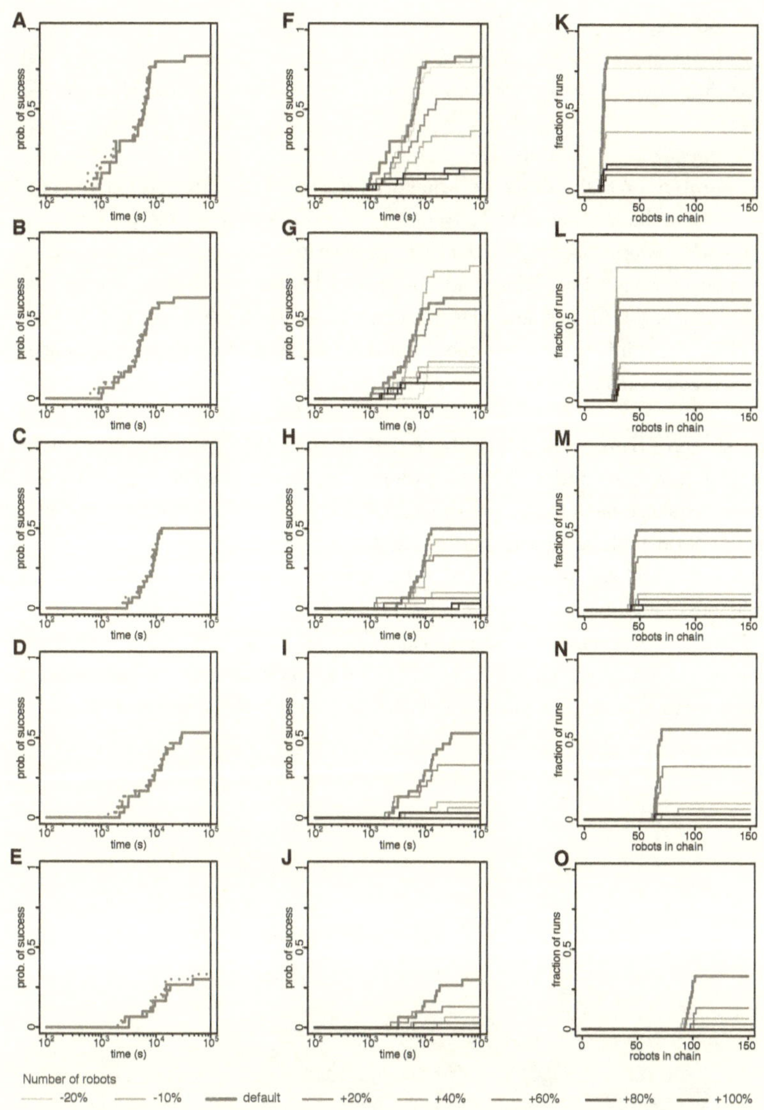

Figure 4.11: **Scalability and robustness of Mark II$_4$.** (**A-E**) Scalability study using the default number of robots in five arenas of different size (see Table 3.4). Empirical run-time distribution for the execution of one (dotted), five (dot-dash), and ten (solid) sequences. (**F-J**) Robustness to the variation of the number of robots between -20% and +100% of the default number (see Table 3.4). Empirical run-time distribution for the execution of ten sequences. (**K-O**) Empirical distribution of the number of robots in the chain as a function of the total number of robots. Arena's area: (**A**, **F**, **K**) 2.10 m^2. (**B**, **G**, **L**) 4.21 m^2. (**C**, **H**, **M**) 8.42 m^2. (**D**, **I**, **N**) 16.84 m^2. (**E**, **J**, **O**) 33.67 m^2.

the leader-election process.

Possible improvement: we could adopt locally-unique identifiers, which have been successfully demonstrated with a swarm of one thousand robots (Rubenstein et al., 2014b).

The number of tasks must be known at design time. All variants of TS-Swarm assume that the number of tasks to be sequenced is known at design time.

Possible improvement: we could let the swarm determine the number of tasks autonomously at run time. This would be straightforward in Mark II_m: when the closed-loop chain is established and the guardians order themselves using a leader-election algorithm, the information on the number of tasks discovered by the swarm in the environment is readily available to all the guardians.

The search strategy in Mark II_m is sub-optimal. All runners execute the same candidate sequence during the exploration of the tree of possible sequences. While the correct sequence has not yet been identified, this approach is not optimal, as only the first runner completes the current candidate sequence and receives a feedback. At that point, if the feedback is negative, all other runners abort their execution and transition to the next candidate sequence, guided by the chain.

Possible improvement: Different runners could be guided by the chain to explore different candidate sequences, in parallel. This would leverage the redundancy of the robot swarm to speed up considerably the exploration of the search space.

Chapter 5

Conclusions

In this **book** we have presented TS-Swarm, a robot swarm with the ability of autonomously executing a set of tasks in the right order, when the right order is unknown at design time.

We began by analyzing the literature in depth and by acknowledging that: i) a general methodology for designing the individual behavior of the robots starting from the swarm-level requirements does not exist yet, ii) most of previous work focuses on the emergence, at the swarm level, of simple mechanical/geometrical properties, and iii) most of the robot swarms demonstrated so far operate under the assumption that the sequence of tasks that the individuals should perform is known (or is deducible) at design time.

We then argued that, for being truly autonomous and operating in the real world, robot swarms should posses more complex cognitive abilities. Among these abilities, a prominent one is autonomous task sequencing—that is, the ability to autonomously find out, at operation time, the correct order of execution of a set of tasks to accomplish the goal mission. With this ability, a robot swarm would be able to operate even when the correct order of execution is unknown at design time. The task-sequencing ability would emerge at the swarm level out of the interactions between robots, which individually would act by reacting to contingencies without being aware of the task-sequencing problem being solved.

To develop a robot swarm with the ability to sequence tasks, we started from observing that a few works from the literature could be associated with a form of path planning. In these works, a sub-group of the robots of the swarm align in space to create a chain-like structure. The robots of the chain landmark the space and act as waypoints for other robots that need to navigate from one end of the chain to the other. For the first time in this **book**, we have acknowledged chaining as a method of path planning and we have performed a step further: in TS-Swarm, we generalize chaining to planning task sequences. In fact, the chain formed by a sub-group of the robots of TS-Swarm fulfills two functions: 1. it assists the navigation between the relevant areas where tasks are to be performed and 2. it identifies/encodes the order in which tasks must be performed. Thus, in a sense, the chain members of TS-Swarm "landmark" the abstract space of the tasks, creating a precedence relation between the tasks themselves. By following the information provided by the chain, other robots can navigate between the areas where tasks are to be performed and perform those tasks in the order encoded.

We have detailed the implementation of two versions of TS-Swarm, which differ from each other for the underlying assumptions: Mark I assumes that, as soon as it performs a task, a robot becomes immediately aware of whether the task was performed in the correct order. Through a large number of experiments with robots and in simulation, we demonstrated that Mark I can successfully sequence three and four tasks. We then made the sequencing problem harder and introduced a new version of the system, which is able to address it: Mark II. Mark II assumes that a robot must perform an entire sequence of tasks before becoming aware of whether that sequence is correct or not. Because of this lack of immediate feedback, the problem solved by Mark II is more complex than that solve by Mark I. In particular, the swarm in Mark II must search the correct order of execution by traversing the tree of all possible permutations of the m tasks. The computational complexity of the problem is thus $O(m!)$. Also for Mark II, we ran a large number of experiments to demonstrate that it is able to correctly sequence three and four tasks. Finally, we showed that both Mark I and Mark II are scalable to the number of robots and the size of the environment in which the system must operate. Both systems are also robust to the variation of the total number of robots that the swarm comprises. After presenting Mark I and Mark II and analyzing their results in the experiments we performed, we discussed a set of possible improvements and extensions that could optimize the performance of each one of them.

Another important contribution of this work is the development of an integrated infrastructure for performing experiments in swarm robotics. The infrastructure includes the hardware and software platforms on which we then implemented TS-Swarm. The infrastructure provides researchers with the tools for implementing their solutions in a simulated environment faithful to reality and then port those solutions smoothly to the real devices. The infrastructure was made available to the swarm robotics research community.

Besides representing an important step towards fully autonomous robot swarms that are able to operate in the real world, TS-Swarm provides a new perspective on one of the historical debates in artificial intelligence: deliberative *versus* reactive approach. On the one hand, in the deliberative approach, robots are designed as planning machines that create a course of actions by reasoning on a model. On the other hand, in the reactive approach, robots simply react to contingencies. TS-Swarm combines the two approaches in a novel way: a simple form of planning (task sequencing) emerges at the collective level from the interactions of reactive robots. Thus, the individuals themselves preserve the simplicity and agility typical of reactive systems, while the swarm as a whole displays abilities that are typical of planning systems. These complex cognitive abilities can endow the robot swarm with unprecedented autonomy and adaptability.

The process of learning task sequences by TS-Swarm bears some resemblance to other learning processes described in the multi-robot and multi-agent literature (Haynes and Sen, 1996; Sałustowicz et al., 1998; Quinn et al., 2003). TS-Swarm learns the correct sequence based on binary rewards: failures and successes experienced after performing tasks. No example of correct behavior is provided to the robots. In this sense, the learning process performed by TS-Swarm can be classified as reinforcement learning (Sutton and Barto, 1998). More precisely, as no value function is explicitly learned (Sutton, 1988; Watkins and Dayan, 1988), the learning process of TS-Swarm could be seen as a form of direct policy search (Baird and Moore, 1999; Baxter and Bartlett, 2000; Anderson, 2000; Rosenstein and Barto, 2001). In Mark I$_m$, feedback is received immediately after the execution of each single task. On the other hand, in Mark II$_m$ feedback is delayed and is received only after the execution of a complete sequence. As a result, the sequencing problem presents a combinatorial nature: the resulting learning process is much more challenging. The robots learn collectively the correct sequence and the path to reach the areas where the task must be performed. A single learning process takes place, as opposed to collective systems in which each agent/robot learns individually a behavior. In this sense, we can qualify the learning process of TS-Swarm as team learning (Panait and Luke, 2005; Buşoniu et al., 2008). More precisely, as the behavior that is collectively learned is the same for all robots—the behavior that each robot (runner) must execute to perform the same correct sequence— the learning process can be qualified as a form of homogeneous team learning (Haynes and Sen, 1996; Sałustowicz et al., 1998; Quinn et al., 2003). Nonetheless, TS-Swarm differs from typical team learning systems (Stone and Veloso, 2000; Parker, 2012; Girard and Emami, 2015) in the fact that the single learning entity is indeed the swarm as a whole, which has an immaterial and distributed nature: the robots operate in an independent manner and no central entity exists that performs the learning process having a global view of the state of the system. Learning takes place at the collective level of the swarm: it is the swarm as a whole that searches the space of possible solutions. Moreover, once the correct solution is identified, the policy to produce is eventually encoded by the chain in a collective and distributed way: each guardian stores the part of policy that concerns the execution of its guarded task. Each runner implements the policy encoded by the chain on the basis of its own state, which is defined by the number of tasks performed and by which guardian is in its proximity, if any.

The system presented in this **book** offers several potential directions for future work. One direction is to improve the performance on task sequencing and test more complex scenarios, such as scenarios with larger numbers of tasks, environments with obstacles or non-convex environments. Another interesting direction would be to empirically study the behavior of the system more deeply and identify some critical parameters—e.g., the minimum number of robots required to successfully accomplish the mission given the size of the environment. Creating a mathematical model that describes the behavior of the system would help proving and ensuring its convergence and other properties— e.g., time of convergence. Finally, the chain formation algorithm could be improved and other types of physical formations could be tested as a base for the distributed task-sequencing algorithm: for instance, chains of moving robots could speed up the execution of tasks as none of the robots would only act as a landmark, lattice or grid formations could improve the exploration of the environment.

This work is a first attempt to endow robot swarms with cognitive abilities that allow them to formulate and adapt their strategies of action online, at operation time. These abilities can contribute to making robot swarms fully autonomous and able to operate in dynamic environments, which are crucial requirements in several real-world applications. We believe this work could open the way to new studies in this direction and promote the development of systems with more complex cognitive abilities, such as more advanced forms of planning and scheduling of tasks.

Appendix A

Appendix

A.1 Introduction

To conduct the research described in this **book**, we developed an infrastructure that provides researchers with an integrated environment for swarm robotics experiments. The infrastructure includes both a simulation framework in which researchers can create and develop their experiments and the physical devices on which those experiments are finally deployed.

When designing a swarm robotics system, the main problem that researchers must face is the definition of individual rules that will result in the expected collective behavior. As deriving the individual rules from the collective requirements is difficult and time-consuming, researchers generally use simulation tools to perform this task. The design of the appropriate control software for robots leverages a simulated environment through a trial-and-error process or through automatic design techniques (see Section 2.2). One of the most widespread swarm robotics simulators is ARGoS (Pinciroli et al., 2012). ARGoS offers a modular architecture that is easy to extend and a really efficient simulation, even with large groups of robots.

Once the user is satisfied with the results obtained in the simulated environment, the goal is finally to deploy the solution on the robots with little effort. ARGoS provides the tools to upload and execute the same control software directly on the robots.

The infrastructure developed for this **book** is composed by the e-puck robot (Mondada et al., 2009) and a device for task abstraction called TAM (Brutschy et al., 2015). The two platforms have been designed, customized, programmed, and integrated in the simulation of ARGoS. The infrastructure was made available to the community so that other researchers can leverage this fully integrated environment for the development of swarm robotics experiments.

A.2 E-puck

The e-puck (Mondada et al., 2009) is a small wheeled robot developed as an open
tool for education and research purposes. Its main advantage over the competitors is
the relatively low cost and thus a broad community of users. The base version of the
e-puck features a limited set of sensors and actuators. Around the circular body of
the robot 8 infra-red transceivers are positioned to perceive the presence of obstacles
or the intensity of the environmental light. Other sensors are a color camera at the
front of the robot, a microphone and a 3-axis accelerometer. The actuators of the
base e-puck are the motors of the two wheels, which can be set to produce a speed
between 0 and 18 cm/s, a ring of 8 red LEDs and a speaker. Additionally, the e-puck
features a bright LED at the front. Given the very limited capabilities of the basic model,
the e-pucks can be enhanced with extension boards. In this **book**, each e-puck features a
ground sensor, which allows the robots to perceive the gray-scale color of the ground on
which they navigate, a range and bearing device that enables local communications
between robots (see Section A.3), an omni-directional camera that provides a 360° view
of the surroundings, and an embedded computer running Linux that enhances the
computational power. The Linux extension board[1] features an ARM[2] processor and adds
all the potentials of a computer running Linux, among which the possibility of using a
USB dongle to connect to a Wifi network and the opportunity to add an additional
board that integrates 3 RGB LEDs. Figure A.1 shows the final configuration of the e-
puck. The complete software model of the e-puck for ARGoS can be downloaded from
`https://github.com/demiurge-project/argos3-epuck`.

A.2.1 E-puck firmware architecture

The addition of extension devices required a re-design of the software running on the e-
pucks. Because of the hardware configuration, the sensors and actuators included in the
basic model of the e-puck are controlled by a PIC micro-controller (dsPIC 30F6014A),
which is programmed in the C language, while the additional sensors and actuators
can be controlled directly from the Linux embedded computer. The ARM processor of
the Linux board can access the ground sensor and the range and bearing via I2C serial
bus, while the omni-directional camera is accessed as a USB device. The PIC and the
ARM processor are connected through a UART serial bus.

The goal of the re-design was to create a completely integrated infrastructure with
ARGoS. Thanks to this infrastructure, researchers can develop control software for

Figure A.1: **E-puck extended with range and bearing, Linux extension board and omni-directional camera.**

e-pucks in a simulated environment and then execute the same code, without modifications, on the physical robots. For this reason the software running on the Linux embedded computer was written in C++, which is the language used to program the core of ARGoS as well as the control software for robots. Another requirement in the design of the e-puck software architecture was that, just like in ARGoS, the control of the robot must be performed in a loop composed of three phases, executed in sequence: *sense, control,* and *act.* In the sense phase the sensors' readings are gathered and made available for the control phase, when a step of the control logic defined by the researcher is executed. The result of the control phase is a set of new values for the robot actuators, which are finally maneuvered in the act phase.

The resulting software architecture is shown in Figure A.2. Its implementation can be found in the real-robot package of the e-puck model for ARGoS.

The software architecture is divided in two main components: the low-level loop (written in C) executed on the PIC of the e-puck, and the main loop (real e-puck main written in C++) that encapsulates the user-defined control software and is executed on the ARM processor of the Linux board. The two software components communicate with each other through a UART serial link. A complete cycle of control is composed of these steps:

1. The Linux board reads the sensors directly accessible (omni-directional camera, range and bearing sensor, ground sensor).

2. The PIC reads the value of every sensor directly connected (8 proximity sensors/-light sensors, microphone, accelerometer).

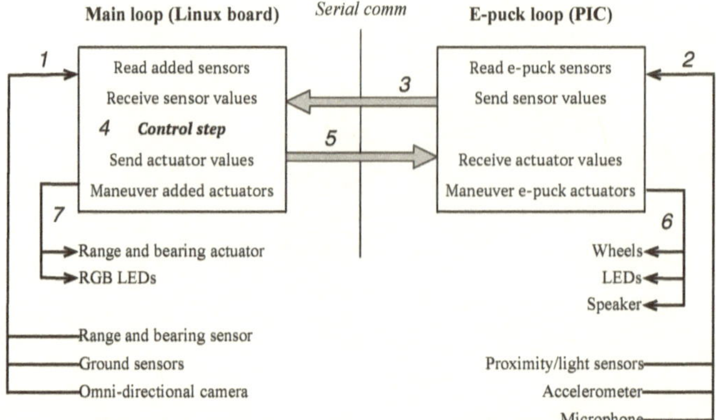

Figure A.2: **Steps of a cycle of control of the e-puck software architecture.**

3. The PIC sends the read sensor values to the Linux board through the serial link.

4. Once the values of every sensor is available, the control passes to the control step defined by the user. Depending on input values and the logic defined in the control step, new values for the actuators are set.

5. The Linux board communicates back to the PIC the new values for maneuvering the actuators of the basic e-puck (wheels, 8 red LEDs, speaker).

6. The PIC maneuvers the robot according to the new values.

7. The Linux board uses the new values to maneuver the actuators directly connected (range and bearing actuator, RGB LEDs).

These steps are repeated in the main loop every 100 ms, which is the default time step duration in ARGoS. The main loop also sends the start and the end commands to the PIC. By keeping track of the state of the communication, the main loop can also handle and recover possible malfunctions.

A.2.2   E-puck in ARGoS

The hardware and software details described in the previous section have been made completely transparent to the user, who must simply use an abstract control interface to access sensors and actuators. The same control interface is implemented by both the simulated and the real robot packages.

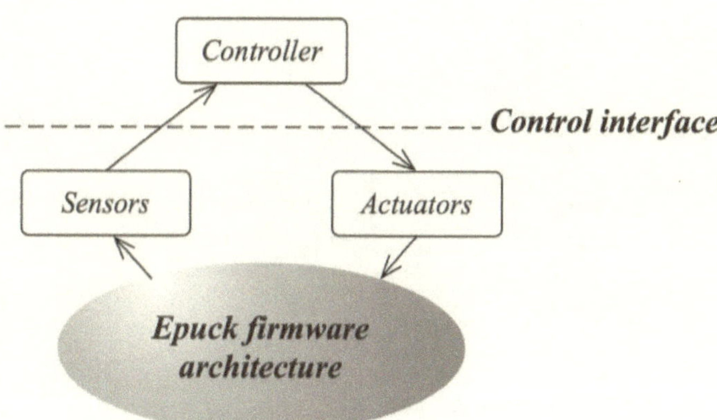

Figure A.3: **The architecture of the real e-puck package integrated in ARGoS.**

Real-robot package In the real e-puck package the control interface is implemented to
hide the low-level details of the hardware and make them transparent to the user. Each
sensor and actuator has its own implementation. In particular, sensors can be divided
in two categories: the ones directly accessible from the Linux board and the ones
accessible only through the PIC. In the former case, sensors are directly read and the
data can be post-processed before being made available to the user control software (e.g.
the images acquired from the omni-directional camera are processed to extract areas of
neighboring pixels in the same range of color spectrum, these areas are aggregated in
structures called blobs, and the blobs are provided to the user with information about
relative position, size and color). For the latter type of sensors the data is acquired
by the PIC and then insert in a data structure (real epuck base) that is sent back
to the main loop through the serial bus. Once the communication is completed, the
module that implements each sensor can read its own data from the received structure.
Figure A.3 shows the e-puck software architecture integrate in ARGoS.

The real e-puck package requires to be compiled for the specific hardware architecture before being transferred on the robot from a personal computer. Information and
a step-by-step installation guide can be found in the documentation folder of the e-puck
model for ARGoS.

Simulation package In the simulation package, the control interface is implemented to
read and modify the simulated 3D space created by ARGoS. The simulated 3D space
is composed by a set of data structures that contains the complete state of the simulation (position and orientation of robots and obstacles, for instance). The state of the

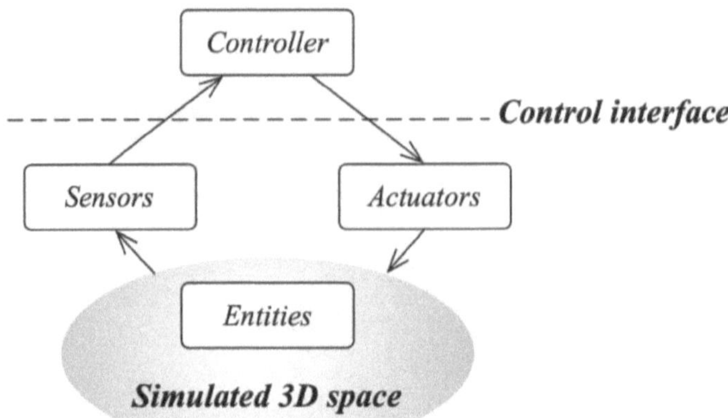

Figure A.4: **The architecture of the e-puck simulation package.**

different functional components of the robots, i.e. the state of the range and bearing device (RAB), is also stored in this set of data structures. ARGoS keeps the simulated 3D space organized in items called entities. Different components' states are stored in different specialized entities (e.g. *RAB-equipped entity*).

Sensors and actuators are plug-ins that access the state of the simulated 3D space. Sensors have read-only access to the space, while actuators can modify it. Sensors and actuators are designed to only access the necessary specialized entities. A diagram of the e-puck simulation package in ARGoS is reported in Figure A.4.

The simulation of sensors and actuators can be tuned by means of several parameters that the user can specify in the ARGoS configuration file of an experiment. Sensors and actuators can be used in an ideal version or in more realistic versions—that is, by adding noise or assigning realistic values to other parameters that modify the behavior of the specific sensor/actuator.

Calibration   One of the most delicate aspect to consider when passing from simulation to the physical robots is the difference between sensors. Every detector is different from the others, sensor readings can be more or less noisy depending on environment conditions and other factors that are impossible to model in a simulated world. For these reasons, users can perform a preliminary step called *calibration* before executing their control software on the physical robots. The goals of the calibration phase are (i) reducing the effect of differences between sensors; (ii) normalizing the raw values given by the sensors in such a way that they are consistent to the ones used in simulation; and (iii) adjusting the ranges of normalization depending on the environmental conditions chosen for the particular experiment.

We implemented a calibration controller for the e-puck model in ARGoS. When given the name of the sensor to calibrate and executed, the controller prints all the instructions that the user must follow in order to proceed with the calibration.

The execution produces a xml calibration file containing the values needed to convert raw sensor readings to calibrated during the experiments. When using a sensor in their experiments, users must specify whether they want to use its calibrated version (by specifying the calibration file produced by the calibration controller) or the raw version. This can be done in the xml configuration file of the experiment.

A.3 Range and bearing

Among the devices of the e-puck, one of the most powerful is certainly the range and bearing (RAB) (Gutiérrez et al., 2009). The range and bearing enables local communication between the robots. The device is equipped with 12 infra-red emitters and 12 infra-red receivers through which it can send and receive messages. Upon reception of a message, the range and bearing is able to measure the relative distance (range) and direction (bearing) of the robot emitter.

A.3.1 Range and bearing firmware

To create the carrier of the emission module, the range and bearing firmware starts a pulse-width modulation (PWM) timer with a period of 1.09 µs. The timer generates an interrupt every 100 µs. The interrupt handler takes the buffered data and sends it to the hardware gates for its transmission. A Manchester code is implemented to allow any data sent at a certain distance to be received with the same intensity by the receiver. The transmission module can be configured in order to send the same or different data from different emitters (single emitters can be disabled as well). Once a transmission request is sent by a master to the device, the communication module decomposes the data for the different emitters with a preamble (6 bits), the payload (16 bits in the original version) and a CRC (4 bits).

The reception module continuously listens if a message arrives. When it detects the preamble of a frame in one of the receivers, the module receives the payload and CRC. If the CRC check passes, the frame is stored in a buffer. If peak detectors of different sensors receive the same signal at the same time, the information is used to calculate the relative direction and distance to the emitter. These two values are also stored in a buffer available to the master.

The interface between the master and the device is given by registers that the

master can write in order to send commands, parameters or data to the board, or read
to receive responses or data back from the board.

The original firmware only supports messages with a fixed payload size of 16 bits.
In many applications this may represent an important constraint, as 16 bits may be too
many or too few. To meet the requirements for this book, and to prevent the need of
implementing costly communication protocols, the range and bearing firmware was
upgraded to support extended and parametric payload sizes. The available payload
sizes are now 8, 16, 24 and 32 bits. The modification involved the extension of the data
structures that store the data to send and receive. Additionally, we implemented a run-
time allocation of this structures and management of the indexes over the structures,
depending on the payload size sent by the master to the board. An additional register
was added to the interface to let the master set the chosen payload size. Four other
registers were added to read or write the two additional bytes of payload.

A.3.2   Range and bearing in ARGoS

The integration of the e-puck range and bearing in ARGoS follows the same principles
described in Section A.2.2. Because of its importance and its many different functions, it
is worth describing both the implementation for the physical robots and the simulation.

Real-robot package   The implementation of the range and bearing device for the
real e-puck had the goal of creating an interface between the user-defined control logic
and the firmware of the device described in the previous section. As the range and
bearing is both an emitter and receiver of messages, both an actuator and a sensor
were implemented.

To realize a bidirectional communication, the user must add the range and bearing in
both *actuators* and *sensors* subtrees of the xml configuration file of their experiment.
Among the parameters that can be assigned to the range and bearing in the xml
configuration file, the *data_size* sets the payload size that the device will use, expressed
in bytes. The value of this parameter must match for the actuator and the sensor and is
sent to the range and bearing firmware at initialization time by writing the designated
register.

To minimize the probability of message loss, the sensor starts a thread whose task
is to regularly poll the range and bearing device asking for new messages received.
Messages are added to a buffer that is read by the main loop once every control step
(along with the distance and the direction of emission). Indeed, the main loop is busy
handling the communication with the e-puck PIC for the most part of the 100 ms cycle
of control. Limiting the reception of messages to the remaining part of the control

cycle would compromise the performance by increasing the number of lost messages. The raw value of range given by the device represents a measure of signal strength between 0 and 4096. In the sensor module, we implemented a conversion to return a more meaningful value, i.e. a value of distance is centimeters. The conversion function, which was derived from experimental data, is reported below:

$$distance = gain * e^{\alpha + (\beta * strength)} \tag{A.1}$$

Where $\alpha = 9.06422$, $\beta = -0.00565$ and $gain = 0.08674$.

Similar considerations can be done regarding the actuator. Once the control logic defines a message to send, the message should be available to the other robots for the longer time-span possible, until a new message is set. Therefore, the actuator should keep invoking the range and bearing for the whole duration of the control cycle (one request to the range and bearing corresponds to a single emission of the message). This was achieved with a separated thread, which requests the emission of the message set by the control logic once every 20 ms (different periods were tested, with shorter periods causing an overload on the I2C bus). The actuator module is also responsible of keeping track of the state and the data assigned to the different emitters. The control logic might perform multiple data-emitter assignments during a control step, sometimes conflicting with each other. The actuator uses only the resulting final state of the emitters and, depending on this state, optimizes the number of operations on the I2C bus to control the range and bearing device. For instance, setting the same payload to each individual emitter would produce 12 write operations on the bus. By recognizing that the same data has been set for all emitters, the actuator module produces instead a single operation on the bus that requests the emission of a payload from all emitters.

Simulation package The simulated range and bearing allows e-pucks to perform situated communication in the simulated environment created by ARGoS. As explained in Section A.2.2, the implementation of sensors and actuators provides an interface between the user-defined control logic and the simulated 3D space of ARGoS. Exactly like on the robots, the simulated range and bearing comprises both a sensor and an actuator. The implementation of the range and bearing is associated to the range-and-bearing medium. In ARGoS, media are entities responsible of dispatching messages or other information from one equipped entity to one or multiple others. To be able to use the range and bearing, it is hence required to add a range-and-bearing medium to the media of the experiment.

Regarding the range-and-bearing actuator, besides the aforementioned data size, it is possible to tune the range of transmission by assigning the desired value, in meters, to the *range* parameter in the configuration file.

The behavior of the sensor is instead more complex and configurable. If the user does not specify any additional parameter, the behavior of the sensor will be ideal: it will receive any message sent by robots in its range at a given time-step and it will be able to measure precisely the value of distance and bearing of each message. To simulate more accurately the behavior of the physical range and bearing device, several parameters are available. The number of messages that can be received in a single control cycle can be limited to the value of the *max_packets* parameter. Gaussian noise can be added to the values of range and bearing, with a standard deviation specified by the *noise_std_dev* parameter. The loss of packets can be simulated, by setting the loss probability to *loss_probability*. Finally, given the very noisy nature of distance measure when using physical robots, it is possible to specify whether to reproduce this noise in simulation by activating *real_range_noise*.

When *real_range_noise* is activated, the measures performed on physical robots are used to produce noise on the value of distance. The procedure adopted was as follows:

1. The distribution of signal strength values was calculated for several fixed distances on physical robots. The distribution of these values was approximated by a log-normal distribution for each distance. In the sensor, a look-up table contains the μ and σ parameters for the log-normal distribution corresponding to each distance. Figure A.5B shows the log-normal distributions for the 9 distances.

2. The parameters μ and σ are calculated by interpolating the actual distance measured in ARGoS with the physical-robot data stored in the look-up table.

3. The simulated value of signal strength is calculated by drawing a number from the log-normal distribution obtained from the previous point.

4. The value of signal strength is converted in distance with Equation A.1. The boxplots in Figure A.5A show the distribution of final sensed distance in function of the actual distance between the robots in the simulation.

Calibration   The calibration of the range and bearing device can be done with the same calibration controller provided in the *testing* folder of the e-puck model for ARGoS. For the time being, only the value of distance read by the sensor can be calibrated. The calibration requires two robots placed at a distance of 20 cm between each other, running the same calibration controller. The robots make an average of the signal strength received over 100 messages. Thanks to this value, they calculate the correct *gain* parameter for the function reported in Equation A.1 when *distance* = 20 cm.

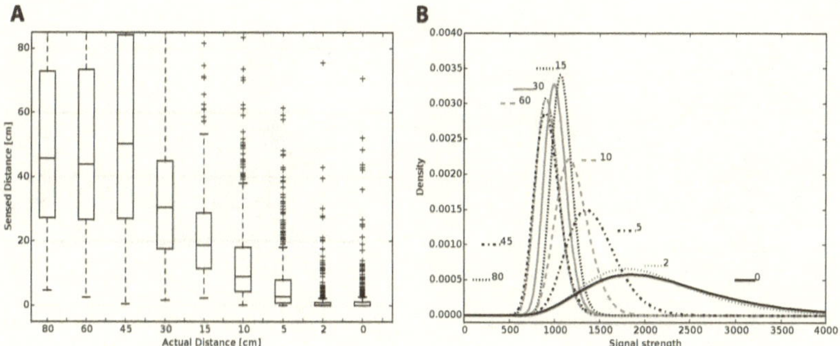

Figure A.5: **Measurements for calculation of noise on range perceived.** In **B**, boxplot distributions of sensed distance against actual distance when real-range noise is active. In **A** the log-normal distributions for 9 fixed distances.

The *gain* obtained will be used to convert (through Equation A.1) the values of signal strength detected by the range and bearing sensor during robot experiments.

A.4 TAM

The infrastructure realized for simulated and physical-robot experiments comprises another device, the task abstraction module (TAM) (Brutschy et al., 2015). The TAM represents single-robot, stationary tasks to be performed by an e-puck. The goal of the TAM is to abstract from details specific to task execution that are not the focus of an experiment. The TAM allows researchers to omit details on task execution and focus on the relevant properties of the tasks such as their logical interrelationships. Complex multi-robot tasks can be abstracted by groups of TAMs. First, a complex task is modeled as the set of its constituent single-robot subtasks and their interrelationships. Second, each single-robot subtask is abstracted by a single TAM and the behavior of the TAMs is coordinated such that it reflects the interrelationships identified by the model.

In the remainder of this section we describe the physical implementation of the TAM, the control framework, and finally the integration with ARGoS.

A.4.1 TAM architecture

A TAM has the shape of a booth, into which an e-puck can enter. The length of every dimension of the TAM is 12 cm. The TAM is equipped with two light barriers, three RGB LEDs, and an IR transceiver for communication.

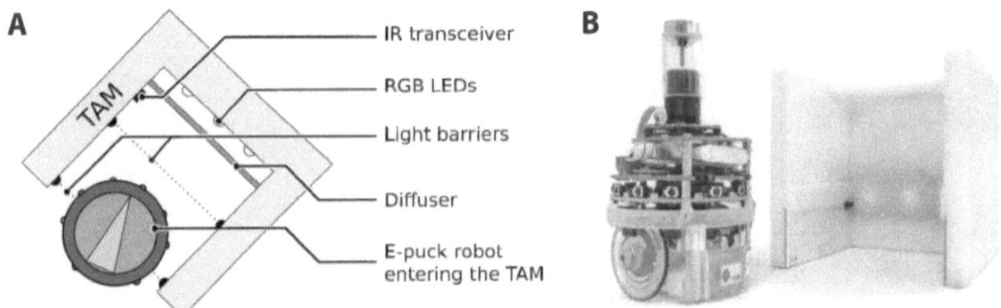

Figure A.6: **Conceptual and real TAM.** The conceptual drawing of the TAM (**A**) and its physical realization next to an e-puck (**B**).

The TAM announces the task it abstracts by using its colored LEDs. An e-puck can perceive the LEDs of a TAM using its omni-directional camera. If the e-puck decides to perform the task represented by the TAM, it moves into the TAM by following the colored LEDs. The TAM can detect the presence of the robot by using its light barriers. Upon detection of the robot, the TAM reacts according to a user-defined logic; for example, by changing the color of its LEDs or by communicating with the robot. Communication between the TAM and the e-puck is realized using the IR transceiver. This communication enables experiments in which the behavior of the TAMs depends on the specific robots working on their tasks. Figure A.6 shows a conceptual drawing of the TAM, and a real TAM with an e-puck robot.

The TAM features an XBee mesh networking module, which allows researchers to synchronize the behavior of multiple TAMs, enabling the representation of complex relationships between tasks and cooperative behaviors between robots. The XBee module can be configured to work on 4 different wireless channels, which allows up to four parallel experiments.

The TAM is open source under the *Creative Commons Attribution-Share-Alike 3.0 Unported License.*

The firmware of the TAM is based on Arduino, an open-source platform using an Atmel AVR micro-controller as central processor. The advantages of Arduino are the wide availability, large community, and relative ease of development compared to other embedded development platforms.

The goal in the design of the TAM's control framework was to create a centralized framework to control groups of TAMs. In this way, the TAM can be remotely controlled by a central computer, it can be used by the computer to gather experimental data, and it operates without being physically connected to the computer.

The control framework of the TAM is composed of two parts: the *firmware* based on

Arduino running on each TAM, and the *coordinator*, running on the central computer. The latter is the software component that handles the wireless communication with all the TAMs connected, keeps the status of every TAM updated, and manages the relationships between the behavior of different TAMs. The *firmware* notifies all events and changes in sensory readings to the *coordinator*, and executes all commands that it receives in return. Commands and notifications are relayed using the wireless mesh network modules of the TAMs. The *coordinator* handles this exchange of commands/notifications and makes it transparent to the user, who can focus on the definition of the logic that controls the behavior of the TAMs.

The software that composes the *coordinator* is programmed in Java to ensure the maximum simplicity and portability. To set up an experiment, the user is required to define two Java classes: a *controller* and an *experiment*. The former is the software that the *coordinator* uses to control a single TAM. One instance of *controller* must be attached to each individual TAM. Similarly to a robot control cycle, the coordinator executes in a loop the *controller* step function, which at every execution takes the state of the TAM as input and produces commands for the same TAM depending on the user-defined logic. In the *experiment*, the user must attach a *controller* to each TAM and handle the logical interrelationships between TAMs.

A.4.2 TAM in ARGoS

The software infrastructure would not be complete without the integration of the TAM in ARGoS. Because the *coordinator* and the control software are programmed in Java, there are some differences between the implementation of the TAM and the implementation of the robots in ARGoS. First, there is no real-robot implementation of the TAM directly in ARGoS, as the TAM's main loop of control is handled by *coordinator*, *experiments* and *controllers*, which are programmed in Java as described in the previous section. Second, to ensure the direct portability of the same control software from simulation to physical TAM, the simulation package provides a C++ wrapper for the Java code used on the physical device.

Simulation package The goal of the simulated package is to wrap the Java software architecture of the real TAM in C++ entities that can be added to the ARGoS simulated 3D space. In this way, the user can create and test an experiment in simulation by implementing control software for e-pucks in C++ and control software for TAMs in Java. The burden of integrating everything in the same simulated world is handled by ARGoS. Once the user is satisfied with the results, the control software for both robots and TAMs can be executed on the real devices without modifications. The architecture

of the TAM simulated package and its interactions with the physical TAM software
architecture are shown in Figure A.7.

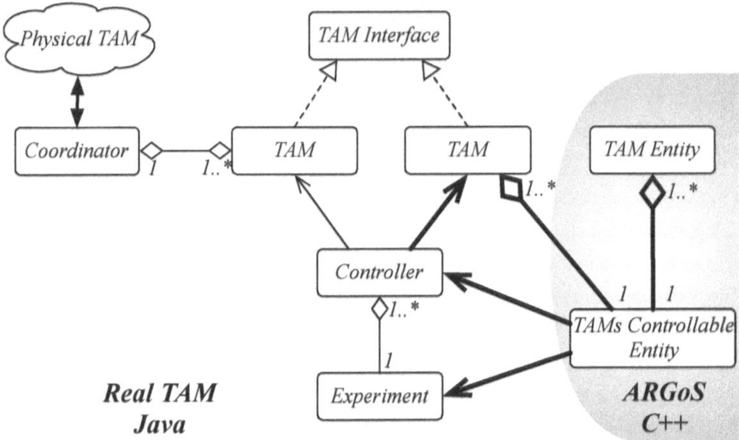

Figure A.7: **TAM software architecture.** On the left side the Java classes for the real
TAM. On the right side the C++ classes in ARGoS. The bold connections are the ones that
realize the C++ wrapper in ARGoS

Two Java classes *TAM* implement the same *TAM interface*, one used for the real
TAM and the other one for the integration in ARGoS. When invoked by the *controller*,
the *TAM* class used for the real TAM, on the left in Figure A.7, relays commands to the
coordinator or reads the state reported by the *coordinator* from the physical TAM. The
implementation of *TAM* for ARGoS, on the other hand, simply keeps the internal state
of the TAM and returns it or modifies it depending on the requests of the *controller*.
The most important class for the integration with ARGoS, i.e. the one the realizes
the wrapper for the Java code in C++, is *TAMs Controllable Entity*. This class is
responsible of creating the *experiment*, instantiating a *TAM* object and a *controller* for
each TAM added by the user to the experiment and binding each instance of *controller*
to the right *TAM*. The interactions between the C++ code of *TAMs controllable entity*
and the Java code of the TAM software infrastructure are realized through the JNI
library. The *TAMs Controllable Entity* is also responsible of managing the timing of
the experiment by calling in a loop the step functions of *experiment* and *controllers*.
Finally, after each iteration, the *TAMs Controllable Entity* takes the state of each
TAM instance (which has just been modified by the step function of its *controller*)
and updates the state of each corresponding *TAM entity* accordingly. The state of the
TAM entities is used to materialize the TAMs in the 3D simulated space of ARGoS.

The effects of events that change the state of a *TAM entity* in the simulation (a robot entering the TAM, for instance) are immediately reported to the corresponding *TAM* object's state, which will be used by the next step function call.

From the user's perspective, all these details are hidden. The user can focus on the implementation of the control logic for the TAMs by defining experiments and controllers.